© Andreas Spaeth 2021

Flightcon Publishing
12 Dukes Court, Bognor Road
Chichester, PO19 8FX, United Kingdom
www.flightcon.net

ISBN: 978-1-9108484-1-8

British Library Cataloguing in Publication Data. A catalogue record for this book is available from the British Library

Cover by Lucy Frankland, Flightcon International
Layout by Lucy Frankland, Flightcon International
Front cover image: Boom Supersonic, Denver CO, USA
Back cover image: Boeing via Geoffrey Thomas, Perth WA, Australia
Inside front cover image: Technik Museum Sinsheim (shutterstock purchased)

Printed in the Czech Republic

Supersonic Travel

Andreas Spaeth

The Author

Andreas Spaeth, born 1966, is one of the most well-known international aviation journalists. He has followed and covered the worldwide commercial aviation industry, including airlines, aircraft manufacturers and airports since the late 1980s. Spaeth contributes regularly in German and English language to print and online media, TV and radio stations and he writes books and records podcasts and video podcasts for outlets in Europe, North America and Australia. The author can be reached via www.andreas-spaeth.net and on Twitter @SpaethFlies

A Note of Thanks

I'd like to express sincere thanks to several people for their valuable support in preparing this book. For expert advice I thank Dr.-Ing. Bernd Liebhardt of DLR German Aerospace Centre. Thanks to the team of Technik Museum Sinsheim, especially Alla Schneider, Hermann Layher and Michael Einkörn. To Alexei Amelyushkin from Moscow, a veteran of the Tu-144. To veteran Concorde pilot Jean-Louis Chatelain and film director Peter Bardehle.

Special thanks to my Australian colleague Geoffrey Thomas who granted me access to his treasure trove of historic Boeing SST vintage photographs. And Jan Altevogt from the Netherlands for making his original photos available. For further photo contributions I thank Ulli Gendrot of Air France, Max Kingsley-Jones and for graphics Jan-Arwed Richter. Jean-Louis Delezenne deserves praise for donating invaluable material.

I am indebted to Boom CEO Blake Scholl who has taken the time for several interviews. And, of course, the many former staff and crewmembers of Air France and British Airways, who enabled me to experience unforgettable moments in and around Concorde and granted me access to places otherwise off limits, including the Concorde jump-seat in flight.

Contents

In scheduled service from 1976 to 2003 – Concorde (AS)

The first supersonic airliner – Tupolev Tu-144 (AA)

Foreword

The Technik Museum Sinsheim houses both supersonic airliners, in front author Andreas Spaeth (AS)

Author Andreas Spaeth on board British Airways Concorde from London to New York, September 2003 (AS)

"The future has to be speed. It seems plausible that someday - at a time near or far Future - a time will come when the average person will have one affordable price for a supersonic flight."

Flight International, May 2020

Supersonic flights have been my most fascinating experience in over 30 years as an aviation journalist. To fly myself on board Concorde had been a big dream since my childhood days in the 1970s, ever since I (at age ten) attended the only-ever appearance of Concorde at my hometown airport in Hamburg, Germany, in April 1976, months after scheduled services had begun. To my luck I was able to take my first Concorde flight as a young aviation reporter in 1993, on Air France from Paris to New York including a helicopter transfer to Manhattan. This was followed by seven further Concorde flights until 2003, including one of the very last ever. This took place on June 24, 2003, from Paris to Karlsruhe-Baden in Germany, where Concorde F-BVFB found a new home at the Auto & Technik Museum Sinsheim near Heidelberg. This is where anyone can marvel at her to this day, spectacularly mounted on a rooftop, next to her former rival, the Soviet-built Tupolev Tu-144.

The most fascinating aspect of flying Concorde each time was the perceived proximity to space, while inside the passenger cabin we were treated to French vintage champagne and caviar (at least on Air France). The pitch-black skies above me, the curvature of the earth below, which could be clearly seen when the atmosphere was clear, proving the earth is indeed a globe. The friction heat made even the inner windowpanes almost painful to touch, while in the cockpit, next to the flight engineer's instrument panel, a gap began to open up after two hours of supersonic cruise. It was wide enough to put your hand inside, or a pilot's cap, and was proof of how the fuselage expanded by almost 30 centimetres on each flight, before retracting again at lower altitude. The incredible take-off thrust with activated afterburners, or the slight kick in the back later, when the fire-spitting afterburners pushed Concorde through the sound barrier. Also riding on the jump seat in the cockpit, most amazing when witnessing the lowering of the visor and droop nose before landing. These are unforgettable memories unique to Concorde.

The people I got to know in and around Concorde have always been at least as impressive. Working on and with this aircraft were the crème de la crème of crews and staff of Air France and British Airways. Everyone was brilliant, in his or her own way. I was particularly delighted to reunite with former Air France Concorde captain Jean-Louis Chatelain in researching this book. It was him who flew me to Karlsruhe on Fox Bravo in 2003. Thanks to my previous flights on Concorde I got to spend precious minutes with celebrities that I would never have met in person otherwise: Sir Paul McCartney, Hugh Grant, Liza Minnelli and Sarah Ferguson, Duchess of York. This will remain equally unforgettable.

There have been hundreds of

Carte d'accès à bord
Boarding pass

Nom du passager / Name of passenger
PAETH/ANDREAS

De/From
CHARLES DE G 2   A

A/To
KARLSRUHE-BADEN

Vol/Flight	Classe	Date	Départ/Time
F 4406	R	24 JUN	1030

Embarquement / Boarding
09H45

Siège / Seat
08A

Porte / Gate   Heure / Time
NB   Poids / Weight

CONCORDE
029

AIR FRANCE

VOL SPECIAL : DON du CONCORDE

AUTO & TECHNIK MUSEUM - SINSHEIM

NOUS SOMMES HEUREUX DE CERTIFIER QUE M
WE ARE PLEASED TO CERTIFY THAT   *Andreas SPAETH*

A PASSÉ LE MUR DU SON À BORD DE CONCORDE LE
BROKE THE SOUND BARRIER ON BOARD THE CONCORDE ON   *24.06.2003*

VOL N°
FLIGHT   *AF 4406*

LE COMMANDANT DE BORD
CAPTAIN   *M. Jean-Louis CHATELAIN*

LE PRÉSIDENT
CHAIRMAN AND C.E.O.

AIR FRANCE

books written about Concorde. This book is not another Concorde book, but one that aims to give the big picture of the phenomenon of civilian supersonic passenger travel. It spans, for the first time, from history to the future. Because besides Concorde, there are other, riveting aspects of history rarely told, such as the fascinating race Western Europeans and Soviets held to be first in the air with a supersonic airliner in the 1960s. And the incredible saga around the American SST project – which at the end of the programme in 1971 became the most expensive aircraft that was never built. In the early 2020s, the phenomenon of supersonic travel has gained momentum again, this time pushed by digital-age private entrepreneurs daring to pull off the dawn of the next supersonic era. And with the first privately built supersonic jet ever, a demonstration aircraft, ready to start testing, this has never been more real since the Concorde days. These new conquests still face extreme uncertainties, so personally I won't want to bet that there will be in fact a new era of supersonic passenger air travel during my lifetime. But I very much hope that anyone who likes to and wants to afford it will be expressing himself or herself to Mach 2 or more again, maybe within the next decade or two.

Andres Spaeth
Hamburg, Germany, April 2021

one

A short history of flying faster than Mach 1

"These so-called Mach appearances, which I experienced as the first pilot, were the initial knocks at the sound barrier."

Heini Dittmar, German Luftwaffe test pilot, 1941

Lockheed SR 71 Blackbird.
[Armstrong Flight Research Center of the United States]

The Messerschmitt Me262 was the world's first jet built in serial production between 1943 and 1945 (source unknown)

Most frequent fliers have at least been very close to supersonic travel, but probably without even noticing it. Even ordinary airliners are able to travel this fast from time to time, in specific conditions of altitude, wind, temperature and air pressure. This happened even in the very early days of the jet era on some flights with the McDonnell Douglas DC-8, in service since 1959, or on the Convair Coronado CV-990 of 1962, which was the fastest subsonic jetliner ever built with up to 1130 km/h cruise speed. In 1961, test pilots brought a DC-8-43 in a dive from 15,800 metres and achieved 16 seconds of Mach 1.01, supersonic speed. The aircraft were over-motorised then, being built very robustly and thus under specific conditions sometimes flew faster than planned. Some areas on the leading edges of wings and rudders of a passenger jet even regularly experience a surrounding supersonic airflow.

Aircraft speed in relation to the speed of sound is always measured according to the surrounding air. If air travels at high speed, for example in the jetstream, and aircraft therefore reach a high overall speed above ground, this is still not supersonic. Wide-body airliners such as the Boeing types 747, 777 and 787 or the Airbus models A380 and A350 often reach cruise speeds of above 1000 km/h, the A380 for instance is designed for up to 1087 km/h. How close this gets to supersonic speed is determined by environmental conditions Decisive is the relation of specific warmth, specific gas constant and thermodynamic air temperature. In dry air and a temperature of 15°C, the speed of sound is 1225 km/h. Above 11,000 m flight altitude, and because of the cold conditions there, it is only 1062 km/h at minus 56°C. Aircraft are designed to reach Mach speeds, not a specific number of km/h. Mach 0.85 for the A380 equals exactly 903 km/h at a cruise altitude of 11,000 m. At sea level 0.85 times the speed of sound would be much higher in km/h, but nobody is flying as fast at this level as no aircraft would structurally withstand the resulting forces. Flight physics has made itself dimension-less by measuring (super)-sonic speeds in Mach numbers, naming this parameter after its Austrian discoverer, the physicist professor Ernst Mach (1838-1916). The unit for simple sonic speed has been defined as Mach 1, anything beyond this is moving with supersonic speed. Modern airliners usually reach cruise speeds of between Mach 0.82 and 0.87.

To build faster aircraft flying higher and further has always been an aspiration of engineers in aerospace. All innovations in civil aircraft production have always come from the military aircraft industry. The very first thing built by humans, that at least in its tip moved with supersonic speeds, was the whip, which was theoretically described for the first time by physicists in 1927. Only in 2008 were scientists able to prove that the tip of the whip reaches twice the speed of sound at the crack. The point in time when mankind itself approached the sound barrier was not reached until the 1940s. It was fuelled by Nazi military flight test research in Germany and their hope for a "Wunderwaffe" (miracle weapon), which was supposed to turn around the Reich's war fortunes. On July 1st, 1941 German test pilot Heini Dittmar reached 1004 km/h with the rocket aircraft Messerschmitt Me-163 near the Luftwaffe test facility at Peenemünde West on the Baltic Sea. It was already known then that from about Mach 0.8 and higher, a phenomenon called buffeting changed aircraft behaviour as supersonic airstreams appeared on the wings. This was also experienced by Heini Dittmar: the aircraft became uncontrollable and was tumbling through the air, steering unresponsive., in danger of disintegrating "These so-called Mach appearances, which I experienced as the first pilot, were the first knocks at the sound barrier", reported Dittmar. Only at the last minute did he succeed in stabilizing his aircraft.

In Great Britain, experiments with combat aircraft were held in 1943 in extreme situations to measure the impact of stresses endured by the pilots. From an altitude of 12.000 metres, propeller aircraft dived down gaining speeds of up to Mach 0.9. Higher speeds were impossible to achieve, as the propellers then generated more drag than propulsion. Due to secrecy and inconsistency of documentation during the war it was impossible to prove, but many years later, veteran engineers then involved stressed that the Germans during their flight tests in the war years assume they broke the barrier of sound four to six times.

Only one case is more concretely known: On April 9th, 1945, very close to the end of the war and German defeat, the 24 year-old ensign Hans Guido Mutke was supposed to train high altitude flying on a Messerschmitt Me-262, the world's first jet aircraft serially built since 1943, in Lagerlechfeld near Augsburg in Bavaria. Therefore he reached 11,000 metres of altitude. To come to assist a colleague under attack in aerial combat, he tilted the aircraft over the left wing and sped downwards under full thrust in an angle of 40°. This caused the Me-262 to get out of control. "The aircraft was rumbling and vibrating, I banged my head into the cabin ceiling", Mutke told a journalist in 2001. He lost 8,000 metres in altitude before he kicked the rudder hard, and thus regained control.

After landing it could be seen that the aircraft structure was

Chuck Yeager reached Mach 1.06 (1,079 km/h) with the rocket-engine powered Bell X-1 in the world's horizontal supersonic flight on October 14th, 1947 (NASA)

seriously damaged. Mutke himself assumes he flew supersonic for about seven seconds. While other war pilot veterans deem this to be impossible, todays scientists at least confirm that the Me-262 was just about capable of reaching supersonic territory. The text of an American pilot manual of January 1946, meant to describe the handling of Me 262s seized in Germany, seems to support Mutke's assertion. "Even in a flat dive of 20° to 30° to the horizon", it says, the Me 262 would reach a speed of 950 km/h. At around 1000 km/h, the controls fail. The manual goes on: "Once supersonic speed is achieved, is reported, these conditions disappear and normal controls are regained."

Only in 1947 was it the Americans' turn, and for more than half a century they have counted it as their pioneering achievement to have been the first to breach the sound barrier. On October 1st, 1947, this was done by a pilot of the US Air Force in a prototype of the North American F-86 Sabre in a 40° dive. The speed indicator, however, had not been calibrated accordingly and there was also no measurement from the ground, so the record was not officially recognised. This was only achieved by

William "Pete" Knight holds a world record still standing today: On October 3rd, 1967 he reached Mach 6.7 (7.274 km/h) with the hypersonic rocket plane North American X-15 (NASA)

24-year-old Charles Elwood "Chuck" Yeager (1923-2020), who eternally engraved his name to the aviation hall of fame on October 14th, 1947. On that day he became the first human to officially break the sound barrier in a Bell X-1 rocket aircraft, essentially a bullet with wings. The aircraft had been specifically designed to break the sound barrier in horizontal flight for the first time, not just in dives as before. It was not even ten metres long and was painted in glowing pink for better visibility in the air - or in the worst case, after an accident, on the ground. The design was simple and didn't even provide an ejection seat, unthinkable in military testing today. Yeager nicknamed his aircraft "Glamorous Glennis" to honour his wife, the name was painted underneath the cockpit windscreen. The X-1 was carried by a Boeing B-29 from the Muroc testing range (renamed Edwards Air Force Base in 1950) in the Californian Mojave Desert east of Los Angeles to an altitude of about 5000 metres. The test pilot climbed through the empty bomb-bay into the rocket aircraft, and 20 minutes after take-off, 6000 metres altitude was reached. The X-1 was released and in due distance to the mother aircraft, Yeager ignited the four-chamber rocket motor, a refinement of German rocket technology. The tanks were filled to capacity with 1177 litres of liquid oxygen and 1109 litres of methyl alcohol.

The X-1 rapidly rose to 12,800 metres. The fuselage shape of the aircraft was deliberately modelled after a standard bullet, as one knew this projectile had a stable "flight" at supersonic speeds, while this shape was aerodynamically less desirable for ordinary aircraft. At prior flight tests, Yeager had already to deal with shock waves when getting close to the sound barrier and accordingly with less efficient elevators. This led to a modification of the X-1 so that the rudder was moved by electric actuators instead of muscle power, thus enabling the record flight. At cruise altitude Yeager accelerated the aircraft to Mach 1.06, equivalent to

1079 km/h. On the ground a double thump was audible once the sound barrier was broken above Edwards, the sonic boom that is unavoidable to this day. Already 14 minutes after release and his successful pioneering achievement, Yeager landed the now powerless X-1 in Muroc.

But it was only conceivable to build supersonic aircraft when they were designed with swept wings, helping to observe the so-called area rule. It had been experienced that on coming close to the sound barrier, the compression of air leads to shock waves coming off from different areas of the aircraft. This leads to increased aerodynamic or wave drag, until what is colloquially called the sound barrier has been passed. Beyond it, the drag becomes less again, but still remains higher than at subsonic speeds. The area rule had been first discovered in Germany in 1943/44 during testing for the jet-powered Junkers Ju 287 bomber. The rule had been registered as a patent and re-emerged about a decade later in the US, serving as a fundamental design essential when conceiving the first supersonic fighter jets.

The rule describes the ideal shape of a supersonic body in the transonic speed range from Mach 0.75-Mach 1.2. For higher speeds, there are other specifications for the shape, which have to take the

After the North American XB-70 Valkyrie bomber program was abandoned, the sole remaining prototype was used for testing purposes in civil programs between 1965 and 1969 (NASA)

shockwaves forming a cone according to Ernst Mach into account. If the area rule is not observed, shock waves can be generated, increasing drag so vastly that supersonic speeds might not be reached. According to the area rule, the fuselage cross-section theoretically has to be smaller by the same amount that the wings add to the total area of the diameter. This is the only way to hinder additional shock waves from forming. In fuselages that stretch out long such as Concorde's, this "wasp waist" in the wing area can hardly be spotted. It is clearly visible, however, on the drawings for the last version of a supersonic airliner by Boeing, the 2707-300 of 1970 (see Chapter 3). In military aircraft with short fuselages, this constriction of the diameter is often easily recognizable.

After breaking the sound barrier, the next goal was to achieve twice the speed of sound (Mach 2), as it seemed to be clear now that no particular difficulties would have to be expected above Mach 1. Due to prestige reasons, chiefly the US Navy was very interested in Mach 2. The vehicle of choice was the Douglas D-558-II Skyrocket experimental aircraft. The 13-metre-long single-seater looked like an enlarged X-1. It originally boasted a jet engine, which was replaced by a rocket engine with four combustion chambers and flight tests began in the summer of 1951. The D-558 again was carried to cruise altitude by a modified B29 bomber and released before engine ignition. On November 20th, 1953, the experimental aircraft wrote aviation history: Pilot Scott Crossfield of NACA (predecessor of NASA) achieved Mach 2.005 (equalling 2066 km/h) in slightly orbital flight in an altitude of 21.950 metres. From now on things moved fast, literally. In February 1954, the prototype of the Lockheed F-104 Starfighter took to the air, the first fighter that could regularly reach twice the speed of sound, its maximum speed was 2459 km/h. It had been a development at breakneck speed, fuelled by the cold war at the time: it was less than seven years from first breaking the sound barrier to a serially produced aircraft often reaching Mach 2.

And that was not the end of the supersonic revolution yet; in the early 1950s scientists and engineers had already started preparations to work on aircraft capable of Mach 3 and even higher speeds. But they had become aware of a problem, as reaching Mach 2 already meant an extreme heating of the fuselage nose tips and wing leading edges to up to 100°C. Due to impact pressure, the air cannot escape sufficiently quickly, its compression results in the heating mostly of the protruding parts of the aircraft structure, despite the extremely cold air of the stratosphere at this altitude. Below such strong heating, the usual aluminium alloys ensure the full strength of the airframe, for higher speeds however, material with higher thermal resilience is needed. Therefore, Bell, who had already designed the X-1 to break the sound barrier, now developed a new test-bed called X-2. It was 11,30 metres long and had wings with a 40° sweep, the airframe consisted of a non-corrosive steel alloy. Its rocket engine generated 68 kN of static thrust, more than double than the D-558 achieving just 27 kN.

On September 27th, 1956, Captain Milburn Apt succeeded in performing a record flight. After release from the carrier aircraft, the X-2 reached an altitude of 21,500 metres and a speed of Mach 3.2 (3350 km/h). There was no time to celebrate, however, as the aircraft got out of control only seconds later and broke apart, Apt died instantly. He had flown the X-2 for the first time while achieving the record and apparently had not been acquainted enough with its specific handling characteristics. Even Mach 3 was a capability that was incorporated into regular aircraft relatively quickly, the Soviet MiG-25 of 1964 could achieve three times the speed of sound for brief periods while the reconnaissance aircraft Lockheed SR-71 Blackbird, introduced in 1966, was even operating whole missions with Mach 3. A total of 85% of it consisted of heat-resistant titan. Its maximum speed was Mach 3.2 (3540 km/h), but as long as the inlets for its engine compressors were not heated above 427°C, this could be enhanced to Mach 3.3. One SR-71 pilot even reported to have flown Mach 3.5 over Libya in 1986 briefly to escape a missile. In any case, the SR-71 is the fastest aircraft with air-breathing engines ever built. For reference: Anything above Mach 5 is called hypersonic.

Another important research aircraft was the only remaining prototype of the North American Valkyrie, originally conceived as strategic nuclear bomber. In its 83 test flights between 1965 und 1969, it collected essential data and insights for future civil supersonic airliner projects in a total of 252 flight hours. To enhance pilot vision on approach, the XB-70 was the first aircraft fitted with a droop nose, exactly like later Tu-144 and Concorde would. And like the former, the XB-70 also had small foldable forward canards. The wings of the XB-70 were swept by 65.5°. In flight, the exposed parts of the airframe heated up to 355°C, therefore it was built of non-corrosive steel alloys, titan and honeycomb composites.

October 14th, 2012 was a remarkable day in the history of human supersonic flight. Not only as it marked the 65th anniversary of Chuck Yeager breaking the sound barrier. But also because Austrian extreme sportsman Felix Baumgartner proved on this day that humans can achieve supersonic speeds much more easily - and entirely without engines. In a pressured module, hanging on a helium-filled balloon, the Austrian reached an altitude of 38,969.4 metres above Roswell/New Mexico in the US, from where he jumped. In free-fall, Baumgartner reached a speed of Mach 1.25 (1357,6 km/h) and lost control over his flying position, as expected, but managed to stabilise himself in time. After four minutes and 20 seconds of free fall, Baumgartner activated his parachute as planned at 1585 metres of altitude and landed safely and unharmed five minutes later.

The reconnaissance aircraft Lockheed SR-71 Blackbird, introduced in 1969, could operate whole missions constantly flying at Mach 3 (NASA)

two

Overcoming design challenges for supersonic passenger aircraft

"There is not the slightest doubt that supersonic overflights will be forbidden by all countries."

Swedish Aeronautical Research Institute, 1967

If temperature and humidity are right when an aircraft breaks the sound barrier, vapours in the air condense into a cloud-like white halo (US Navy)

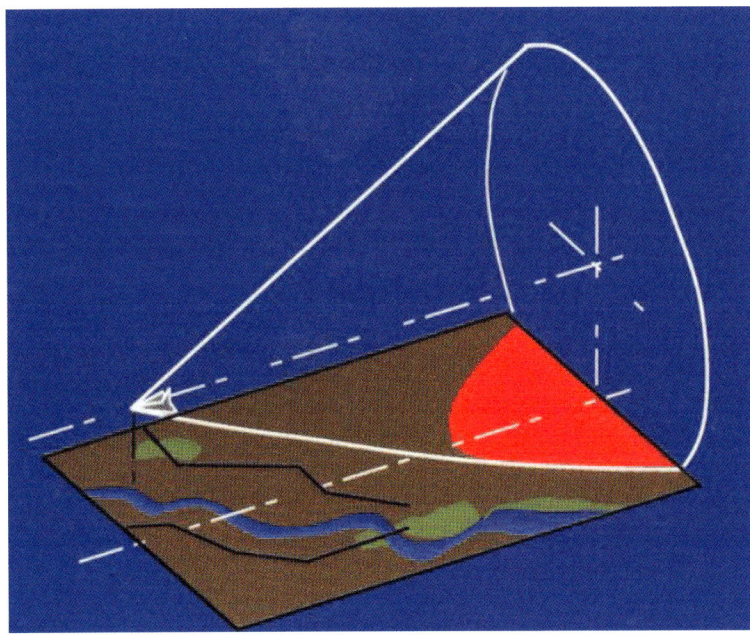

Only in the 1960s it became clear that the sonic boom (seen in red) trails an aircraft on the ground as an inevitable physical phenomenon (AS)

Supersonic flights for military operations, with highly qualified and specially trained pilots on board, are one thing. Especially as aircraft on such missions are only capable of flying supersonic for a very limited period each time, just for minutes on end. It was a totally different technical challenge, however, to conceive regular civil airline passenger traffic in this hardly known frontier that is the upper stratosphere. To achieve the necessary reliability and to enable safe supersonic flying for several hours in a row was a formidable challenge, especially with the limited technical and design tools of the 1950s. But still, a decade after the second world war had ended, the aerospace industries of France, the UK, the USA and the Soviet Union, all the relevant countries at the time, seemed to agree: the future of passenger air transport would be in supersonic flying. This belief continuously reigned until the early 1970s. When Boeing started development of the 747 "Jumbo Jet" from the mid-1960s, this work was considered to be second-rate, as prestige and resources were predominantly being pumped into Boeing's planned supersonic airliner. The 747, it was the assumption, would just bridge the period until supersonic jets would take over passenger flying and the 747 would be relegated to cargo flights. This can be seen in its design – with the cockpit on the upper deck and the famous hump, to enable easy loading through the nose, which could be opened.

All of the big aircraft manufacturers at the time initiated feasibility studies for supersonic airliners – and their results were all the same: within 20 years it should be technically possible and commercially sensible to design such aircraft, manufacture them and put them into service. However, this was mostly wishful thinking, as the studies lacked any sense of realism, as Ronald Davies, an aviation historian, later demonstrated impressively in his pamphlet book *Supersonic (Airliner) Non-Sense*, of 1998 (see Chapter 5). Davies knew what he was talking about, as he himself was part of the process of conceiving such overly rosy studies for the manufacturer de Havilland earlier. He, for example, later proved that the UK only acted to build a supersonic airliner out of fear of losing its global leadership in commercial aircraft manufacturing. Fears were caused by tough earlier experiences during the Comet days, the initially problem-prone first jet airliner, and later with the Vickers Viscount turboprop liner. But now the Americans pressed forward and England was afraid to lose out. "Thus was born the idea of leap-frogging the opposition. The plateau of 600 mph with the jets had been reached. Very well: the British would move on to the next logical step – and at the time it was considered to be not much more than that – and build a supersonic 1,200-mph airliner," Davies wrote. In November 1956, the Supersonic Transport Aircraft Committee (S.T.A.C.) convened for the first time. Its final report was long kept under wraps, it set the expected development costs artificially low at US$375m (today about US$3bn). "The prevailing mood was to 'press on, regardless' as this was Britain's chance to lead the way again. Any suggestions that a supersonic airliner might not be economic to operate were summarily dismissed or ignored as the outpourings of Doubting Thomases," asserted Davies. "The question of the hour wasn't *if* an SST would be built, but simply *when* and by whom, and whether it ought to be Mach 2 or Mach 3." The acronym "SST" for supersonic transport was established in the 1960s and mainly stuck in the US, but is still used today.

The list of problems engineers had to solve on their way to supersonic airliners was formidable. The biggest hurdles involved noise. The fact that supersonic flights would inevitably bring two sources of noise with them was not clear to everyone then. First the intense sound of the engines, especially at take-off with ignited afterburners. At rotation and later to break the sound barrier, the afterburners briefly inject extra fuel into the hot engine exhaust, doubling thrust, noise and fuel consumption at once. And then the sonic boom. "At first, it was not recognised as an insuperable problem. Most of us had become familiar with the double 'boom-boom' as we watched the Hunters at the Farnborough Air Show", recalled Ronald Davies. But engineers had to learn soon, however, that the intensity of the boom increased, the bigger the aircraft was triggering it. A hundred-seat airliner would cause a much louder double boom than a small fighter jet. Only now the understanding sank in that a jet flying beyond Mach 1 is laying a continuous noise carpet as long as the supersonic flight lasts, not just the moment the sound barrier is passed. "Not to worry, the engineers and project managers and aero-dynamicists said. This is simply another technical problem that will be overcome, just as we have solved all the other problems in the past," reported Davies. But there was nothing to solve here. The sonic boom is a basic physical phenomenon that cannot be avoided.

Soon it became clear what that meant: massive

restrictions of supersonic flight profiles at least over land. And correspondingly, a massive impairment of all profitability calculations so far for commercial supersonic flight operations. But this understanding never really found its way into the hopelessly sugar-coated scenarios of profitability. Already since 1955, a UK law called the Air Force Act threatened pilots of the Royal Air Force with fines if they flew their aircraft in a manner that caused unnecessary disturbances on the ground. For the envisioned passenger flights, plans called for a package of measurements: supersonic flying only from 130 km after take-off at the earliest and just up until 130 km before landing at most, and not below 35,000 feet (10,670 metres). From there, aircraft would climb to their maximum cruise altitudes between 60,000 feet (18,230 metres) or in the case of the Boeing 2707 even 70,000 feet (21,300 metres) and accelerate at the same time. The remaining time for cruising at maximum altitude and speed was limited due to the prolonged period necessary for acceleration and deceleration. These intermediate flight phases took as much longer as the higher the final cruise speed would be.

More immediate than the inevitable boom coming from high altitudes was the infernal noise of the supersonic jet engines emitted close to the ground. Due to the aerodynamic shape of supersonic aircraft, their engines were required to deliver substantially more thrust than ordinary ones. The outward shape of a supersonic aircraft, optimised for high-speed cruising, is not made for slow phases of flight and therefore has to be "pushed" into the air with increased propulsion. The objective was that supersonic aircraft were not supposed to emit more engine noise than the first generation of jet airliners such as the Boeing 707, DC-8 or VC-10. Anyone who has not been able to follow these smoky monsters with his or her own ears on take-off from close quarters cannot imagine these days, what kind of physical torture this could mean for a human being. In 2011, the author had the opportunity to witness one of the last VC-10s of the Royal Air Force on take-off – the windows of the building were shattering, its rotation almost resembled a rocket launch. Today's ears are almost spoilt concerning engine noise, so immense has been the progress made in producing quieter engines in the last decades. But in the 1960s, early designs of supersonic airliner engines were far away from measuring up even to noise levels of the early jetliners.

"The engine thrust and the

Breaking the sound barrier or the moment before becomes visible when moisture is trapped within shock waves as seen here, seemingly building a kind of wall (US Navy)

efficiency requirements for a supersonic airliner seemed to be completely at odds with those for an engine offering substantial reductions in noise," wrote Tony Buttler. The noise was generated primarily from the shearing effect of its exhaust gases against the static air of the surrounding atmosphere. No less than 250 different designs of noise-absorbing silencers on engines were tested between 1965 and 1970 by the French aerospace agency ONERA. It conducted tests on scale models and later on Concorde prototypes, including, as Buttler points out, a dizzying array of silencers: corrugated ones mounted on the primary nozzle, those with fluid injection and lobe, blade and spade silencers, all of which were only to be activated during take-off and landing, when the aircraft was closest to the ground. They accelerated the mixing of the hot exhaust gases by sucking in cold ambient air. In the end, the problem was solved with retractable spade silencers which were also used to activate reverse thrust after touchdown. This enabled Concorde to ultimately reach the noise levels of early jetliners – which however were obsolete by the point of entry into service of the supersonic airliner. So the noise of Concorde taking off always remained an anachronism.

The noise caused both by the engines and the sonic boom remained the principle argument of the quickly emerging anti-Concorde, anti-supersonic movement who were behaving more and more militantly. Their slogan was "Ban the boom". Already, in the 1960s, possible damage to the ozone layer as well as cosmic radiation that humans on board could be exposed to came into the focus of critics. This was referred to in an internal report of the UK aviation ministry in June 1965: "A climate of opinion rather hostile to the operation of supersonic air transports has built up in this country. In addition there is the very natural reaction by people against the increasing noisiness of the world, and a tendency to resist any innovation which looks like increasing the general disturbance, while not bringing any obvious advantage to the ordinary citizen"; this was the realistic assessment of author A. E. Woodward-Nutt. This was supported by the Swedish Aeronautical Research Institute, whose director wrote to the British Minister of Aviation in October 1967: "In my opinion there is not the slightest doubt that supersonic overflights of the SST will ultimately be forbidden by all countries".

Besides the noise problem there is also a long list of extreme design challenges for a supersonic airliner. So a powerful pressurised cabin air system needed to be developed to create an atmosphere equalling conditions at 1700 metres above sea level, even at altitudes up to 20,000 metres. A fundamental aerodynamic problem was the shape of the wings, as what is best for cruising at Mach 2 is not suitable for slow flight phases close to the ground. Initially, swing-wings with variable geometry seemed to be an option to adapt aerodynamics, according to the requirements of different phases of flight. But this mechanism proved to be too complex and overly heavy. In the late 1950s there were lots of design ideas, which would appear bizarre today, like a "Project X" concept by Vickers with an M-shaped wing configuration. But soon researchers everywhere came to the same conclusion – the only way to go would be delta wings. These were optimised for supersonic cruise and at the same time offered enough wing area with sufficient volume to integrate big fuel tanks, while maintaining a limited wingspan to induce as little drag as possible.

A crucial question in designing a supersonic aircraft is which material the airframe is made of. The high speed creates substantial friction heat, exposing the airframe to the stress of extreme temperatures. After two hours of cruise at Mach 2.02, the tip of Concorde's droop nose regularly measured 127°C – despite minus 60°C air temperature at this altitude. Other exposed parts of the fuselage were heated to up to 120°C. For even higher speeds, Mach 3, for example, was foreseen for the Boeing 2707, regular aluminium alloys cannot provide the required strength anymore, meaning important parts of Boeing's SST would have had to be produced from titanium. Besides the material itself, the right component design of the airframe was critical. In Concorde's structure, for the first time whole sections were produced from one piece to enhance stiffness. The windows especially, coming in miniature size to ensure structural stability, posed a specific challenge for the durability of the airframe, therefore these sections were milled from one piece each per side of the fuselage. Another peculiarity of supersonic flight is the increase of heaviness on the tip of the aircraft in flight above Mach 1. This requires the transfer of fuel into trim tanks to adjust the centre of gravity. Concorde had such trim tanks in the forward wing area, but for the first time also in the vertical stabiliser. There were also considerations made of how to operate entirely without conventional fuels. This was the objective of a proposal by Sud Aviation in France in March 1958: a nuclear-powered version of its supersonic design called Super Caravelle at the time. Uranium fuel would have provided "almost unlimited range" and solved the problem of the weight and volume of kerosene on board. But in turn this created many new problems: how to cool the reactor behind the cabin, separated by a lead shield, when the fuselage heated up during supersonic flight. The idea was to use heat exchangers. Major operational challenges were adding up, for example the long ground times as both reactor and engines would have only cooled down slowly. Even in the euphoria regarding civilian use of nuclear energy at the time, such hurdles proved to be just too hard to overcome.

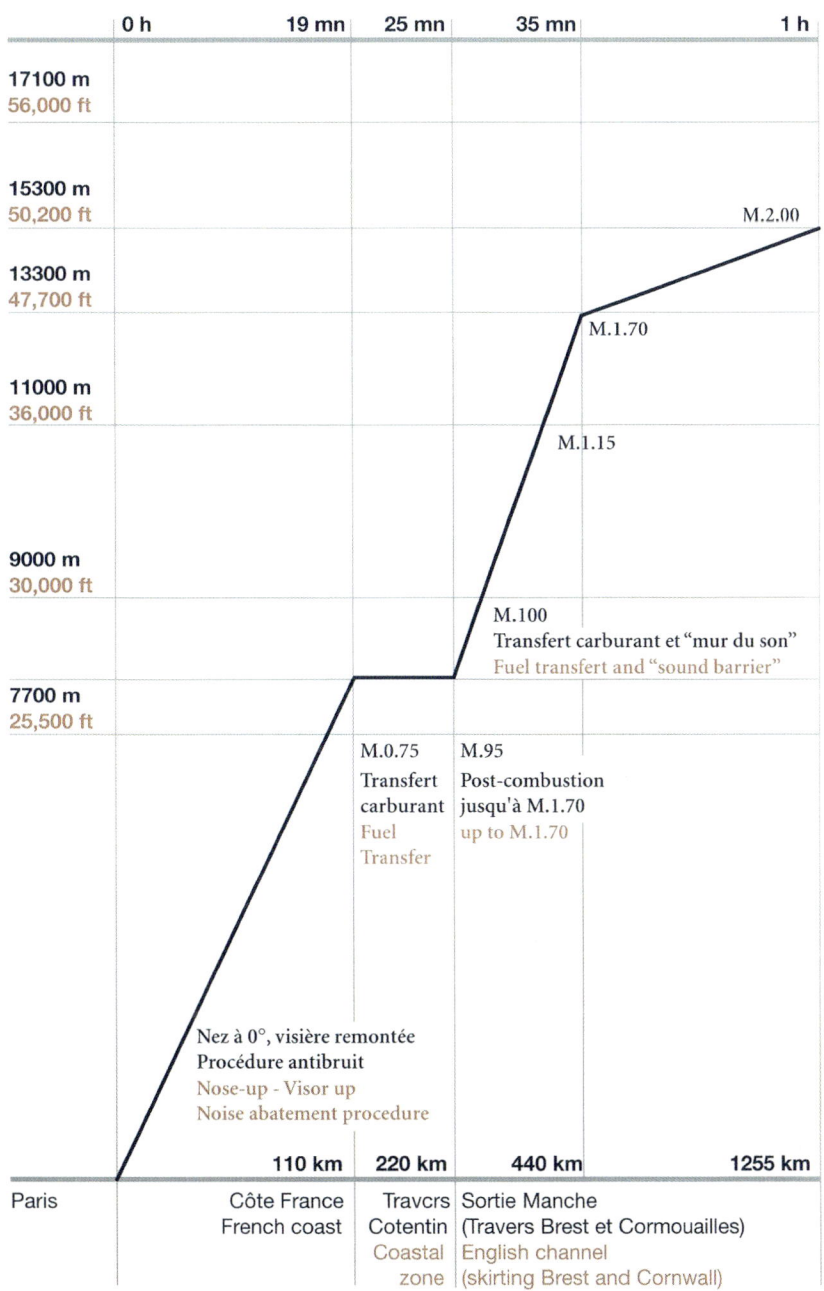

	0 h	19 mn	25 mn	35 mn	1 h

17100 m
56,000 ft

15300 m
50,200 ft

M.2.00

13300 m
47,700 ft

M.1.70

11000 m
36,000 ft

M.1.15

9000 m
30,000 ft

M.100
Transfert carburant et "mur du son"
Fuel transfer and "sound barrier"

7700 m
25,500 ft

M.0.75   M.95
Transfert   Post-combustion
carburant   jusqu'à M.1.70
Fuel   up to M.1.70
Transfer

Nez à 0°, visière remontée
Procédure antibruit
Nose-up - Visor up
Noise abatement procedure

	110 km	220 km	440 km	1255 km

Paris	Côte France	Travcrs	Sortie Manche	
	French coast	Cotentin	(Travers Brest et Cormouailles)	
		Coastal	English channel	
		zone	(skirting Brest and Cornwall)	

Croisière approche et descente

Cruise approach and descent

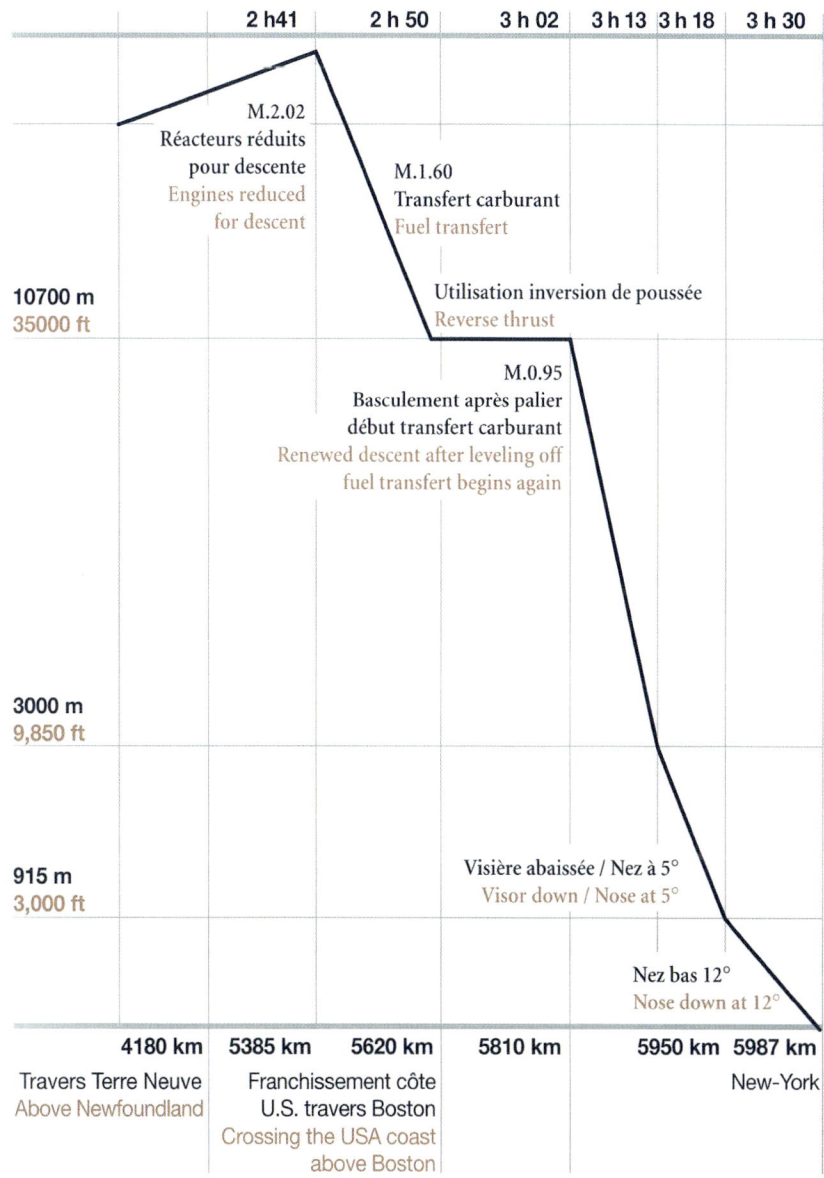

	2 h41	2 h 50	3 h 02	3 h 13	3 h 18	3 h 30

M.2.02
Réacteurs réduits
pour descente
Engines reduced
for descent

M.1.60
Transfert carburant
Fuel transfert

Utilisation inversion de poussée
Reverse thrust

10700 m
35000 ft

M.0.95
Basculement après palier
début transfert carburant
Renewed descent after leveling off
fuel transfert begins again

3000 m
9,850 ft

915 m
3,000 ft

Visière abaissée / Nez à 5°
Visor down / Nose at 5°

Nez bas 12°
Nose down at 12°

	4180 km	5385 km	5620 km	5810 km	5950 km	5987 km

	Travers Terre Neuve	Franchissement côte				New-York
	Above Newfoundland	U.S. travers Boston				
		Crossing the USA coast				
		above Boston				

The profile of a scheduled Concorde flight from Paris to New York shows that after an hour Mach 2 is reached and a maximum of 15,300 meters of altitude (AF)

three

The role of the USA
Boeing 2707 & Lockheed L-2000

"This Government should immediately commence a new program in partnership with private industry to develop at the earliest practical date the prototype of a commercially successful supersonic transport superior to that being built in any other country of the world."

US President John F. Kennedy, June 5th, 1963

Future scenario for supersonic flight operations with Boeing 2707-100. [GT]

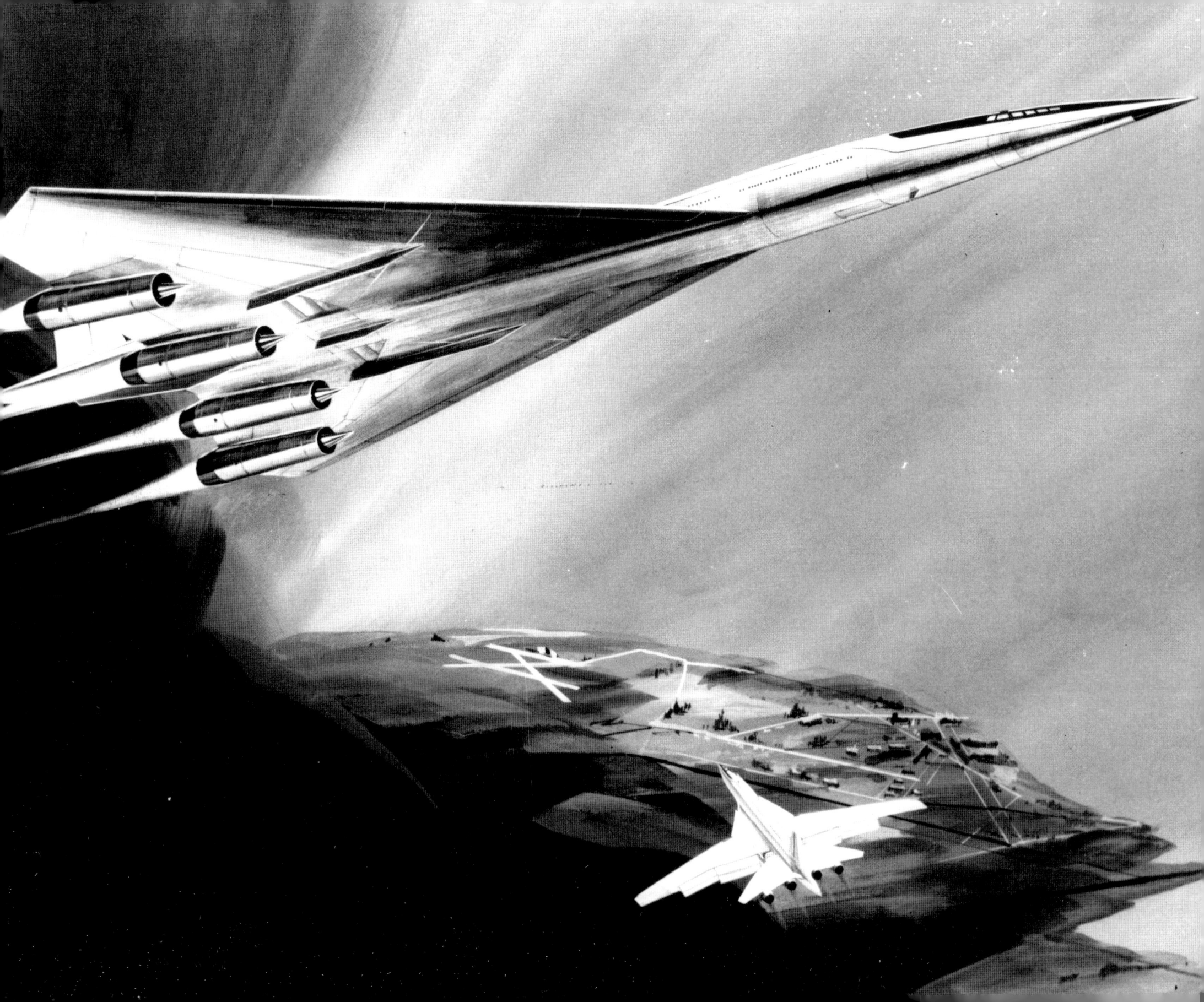

In mid-1966 Lockheed finished work on a wooden mock-up of the planned L-2000, supposed to be 83 meters long, more than all current airliners (GT)

Mock-ups were mostly built for promotional reasons. Here Braniff flight attendants in their harlequin-like uniforms pose on the wing of the L-2000 (GT)

The history of the USA's early 1960s attempts to succeed in the accelerating race to conceive a supersonic airliner has all the elements of a gripping novel: drama, intrigue, political infighting, technological aberrations, flashes of genius, brilliant engineering achievements and mislead market forecasters, triumphs, hopes, nationalism and bitter disappointment. All that kept alive for a whole decade by pumping in amounts of money that still appear outrageous today. At the end there was nothing tangible in return, a horrendous ending became the better option over an otherwise unending odyssey. But let's start at the beginning.

Back in 1952, Boeing began small-scale research work on a possible supersonic airliner. As fighter aircraft were by now constantly breaking the sound barrier, the question was obvious, would this at some time also be an option for a quantum leap in civil passenger transport. Speed is a striking sales argument after all. At the time, there were not even subsonic airliners in the US yet, that only changed with the DC-8 and Boeing 707 in the second half of the 1950s. From 1958, Boeing established a permanent research committee, coming up with the Boeing 733 as the initial study of a supersonic passenger jet. Practical planning started in 1960. An idea was to hang a kind of passenger compartment called "people pod" under a supersonic Convair B-58 Hustler supersonic bomber, the first of its kind in military operations, instead of the rocket-shaped weapon container. It was supposed to be equipped with windows and a pressurised and climate controlled cabin, in which five test persons would be used to evaluate how the human body would cope with supersonic flying. This never happened, and the 1961 study of a 52-seat supersonic airliner based on the B-58 also never made it beyond the drawing board.

At the time, the UK was ahead of the US in supersonic research by two to three years. On January 15th, 1960, the *Washington Post* reported about Soviet designer Andrei Tupolev and his plan to build a supersonic airliner. At the same time, it was foreseeable that there would be a collaboration between France and the UK in building a supersonic airliner; both countries had already done intense research in this field since the 1950s. In the US, the assumption was that both European countries as well as the Soviets were aiming at introducing commercial supersonic operations between 1965 and 1968. The Americans saw themselves under pressure, if they didn't want to risk to be left behind in the emerging race. They were driven by fears that

US airlines could order foreign supersonic airliners and the US could lose its enormous advantage in the global civil airliner arena for decades. At that time, the Americans could still have accepted British advances and teamed up with them in building a supersonic transport (hence the widely used term "SST" in the US ever since). But the US representatives were unable to break the mould in that respect.

"In response to approaches to explore possible collaboration with the UK, all the American manufacturers had been polite but negative and unenthusiastic," writes Tony Buttler. With US President John F. Kennedy taking office in January 1961, the SST subject gained a whole new dynamic in the US. Already by August 1961, Congress had authorised US$11 million for an SST feasibility study, in September the aviation authority FAA established an office for the management of the SST program. That was unusual, as it assigned a key role to this institution in the emerging process, going far beyond the certification and oversight functions that the FAA usually executes. Also the fact that Kennedy had appointed SST enthusiast Najeeb Halaby to head the FAA as Administrator clearly showed the direction in which things were heading. "Halaby successfully fended off British overtures for a joint effort

until the design philosophy of each had hardened, thus making cooperation all but impossible", even the official FAA historian later stated. In October 1962 Congress released another US$20 million for preliminary work on the SST. Still the official start of the Anglo-French Concorde project in November 1962 caught the US somehow off guard and increased fears they would permanently fall behind the Europeans in an important future marketplace. At the same time Halaby was pushing President Kennedy, positive results of the feasibility studies in hand and with the context of the competitive pressure from Europe, to put on a major SST program. The SST advisory group morphed into an SST committee in January 1963, chaired by Vice President Lyndon B. Johnson, showing how much the government already saw the priority of a supersonic airliner at this point.

It is possible to get fascinating insights from 1963 documents and recordings which were publicly released only in 2014. Among them a memorandum of FAA Administrator Halaby to President Kennedy of June 3rd, 1963; this and the following days were crucial in clearing the path for the American SST undertaking. While in Europe, important momentum came from the manufacturers themselves, in the US it was the government getting active and

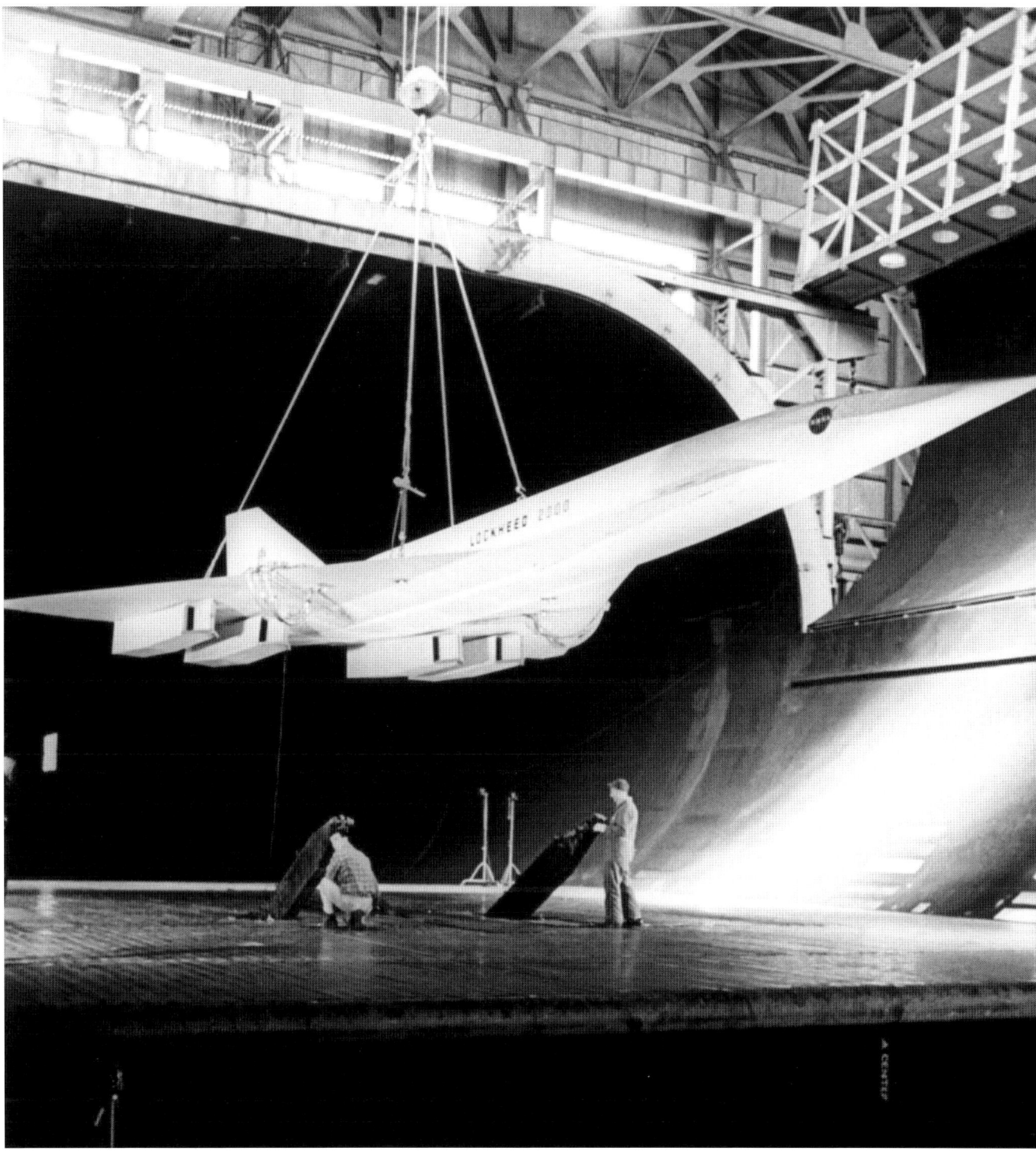

The Lockheed L-2000 was tested as a model in the wind tunnel. Its design was conventional with a double delta wing (GT)

Now in development at Lockheed:
The nation's biggest, fastest Supersonic Transport

In laboratories, wind tunnels, and engineering offices at Lockheed's new Rye Canyon Research Center, one of America's most experienced design teams is hard at work on the nation's biggest, fastest Supersonic Transport in today's race for world transport leadership.

The Lockheed Model 2000 will cruise at 2,000 mph more than 70,000 feet above the earth—yet will land at conventional speed on today's jet runways. The secret of such superior flight performance lies in a new kind of wing. The very large "double-delta" wing of Lockheed's design compensates aerodynamically for the differences between supersonic and subsonic flight—*and does it without movable wing segments.*

For landings, airline pilots agree, the Lockheed design provides the best pilot vision ever offered on a commercial jet. The articulated nose fairing moves downward 15° on approach to the airport *(shown in model above).* The increased visibility allows landing in weather minimums even lower than those permitted today.

Lockheed's 30 years of aircraft experience are behind this new supersonic jet. Its unexcelled speed will produce three to four times the work per day of present jets; its construction is of superstrong, lightweight titanium. Of all the world's plane manufacturers, only Lockheed is building titanium superalloy jet aircraft —such as the spectacular YF-12A Interceptor (announced as the A-11). Proving the superiority of the double-delta wing design, the YF-12A is now flying at speeds and altitudes even greater than those planned for the U.S. Supersonic Transport.

Design, experience, skill, facilities—all are ready *now* at Lockheed to build the plane that will keep America first in air transportation. *Lockheed-California Company, Burbank, California: A Division of Lockheed Aircraft Company.*

LOCKHEED

In this magazine ad from 1964 Lockheed praises the L-2000 as "the nation's biggest, fastest" SST (GT)

sending the manufacturers into the race with taxpayers' money. In his memorandum, Halaby was making suggestions to the President for the content of his major address to be held two days later in front of graduates of the US Air Force Academy. He suggests Kennedy to say:

"We have today decided to commence immediately a program of financial help to the aircraft industry to develop a commercial supersonic transport. Supersonic flight is not new – as all of you here know well. What is new is the challenge of applying this technology to serve the world's civil airlines. We have today accepted that challenge, and expect to send to the Congress shortly a request for $100 million supplemental funds for this purpose.

The cost of developing airplanes – much as the cost of everything else – has gone up. The cost of developing an SST is large – it could be as high as one billion dollars. This is more than our aircraft manufacturers can initially commit to such a difficult and risky program. But we cannot permit this high cost to pre-empt this country from participation in the logical next development of commercial aircraft. The US has always led in the field of long range aircraft – and we want to make sure it continues to do so. (…)

The result must not only surmount all technical obstacles but it must also result in an economically efficient aircraft. We aren't so much concerned with having the first SST – although with dedication and luck we might well have it. What we want is to have the best aircraft, one which transports people and goods safely and swiftly – and at a price that the traveller can afford and the airline finds profitable. If we cannot achieve this, the program will have to be cancelled or deferred until technology permits us to achieve it." (Najeeb Halaby, Memorandum for the President, Subject: The Commercial Supersonic Transport – The Next Steps, June 3rd, 1963, jfklibrary.org)

One day, however, before John F. Kennedy could solemnly speak these words in front of graduates of the US Air Force Academy, the whole undertaking met some turbulence. Nobody had reckoned with the chutzpah of Juan F. Trippe, legendary boss of Pan American World Airways. Trippe announced on the same June 3rd, 1963 in a press release that Pan Am had signed six options for the European Concorde. At the same time, Trippe made it clear that he much rather would order American supersonic airliners if they would be available. But he still opted to secure precautionary options on Concorde with down-payments, but without entering into any obligation to buy. Possibly a deliberate provocation to push the American industry into action. John F. Kennedy, in any case, was outraged at this move and its delicate timing. Everyone can listen to a recording of an original call today on YouTube, in which he lets off steam in talking to his Secretary of the Treasury, Douglas Dillon, after Juan Trippe had rained on his parade. "I think he ought to retract that thing," said Kennedy. "I think you ought to call him up Doug, and say that we're damned sore about this. He knew the United States … my God, I had it in my speech tomorrow," the President was searching for words to express his anger. His main concern was that Pan Am's announcement could shoot down the US SST program, which had not even been launched. And he got back to it: "If he is so indifferent to what the United States government is doing, I think, Doug, you ought to call up and stick it right up his ass. I want him to eat that today, because otherwise we can't possibly go ahead… And I'm really … going to spend our time screwing Pan Am." It came neither to one nor the other; Pan Am only cancelled its Concorde options in 1973 and Kennedy went ahead with launching the American SST program the next day anyway. On the side-lines he was overheard as saying: "We'll

beat that bastard de Gaulle". At the same time, the airline industry was sceptical, with IATA Director General William Hildred putting his rejection this way: "I hope I shall not live to see that damn thing." Najeeb Halaby had already summarised the risks at the end of his memorandum: "Mr. President – the two biggest risks in this program are the US industry may not be able (1) to overcome the sonic boom so that it will be tolerated by the population, and (2) to get the costs down to a competitive level." To then recommend: "These risks appear worth taking."

While several manufacturers had sketched possible SST specifications already in 1960, they now applied less modest specifications, also to clearer distinguish a US SST from the emerging Concorde. Initially one was envisioning an SST carrying 80 to 100 passengers, which, however, was supposed to fly with Mach 2.6 (3220 km/h) at up to 75,000 feet (22.860 metres), higher and faster than Concorde, and was to be exclusively deployed on long-haul routes. Ideas to also fly supersonic on shorter routes had been dismissed already as unrealistic. In Halaby's memorandum of June 3rd, 1963, specifications were defined as follows: Speed higher than Mach 2.2, but below Mach 3. Range at least sufficient for transatlantic services between

Boeing's first design studies for an American SST in 1964 partially resulted from research of the 1950s. First official contender was initially the Boeing 733 (GT)

New York and Paris. Seating for 150 to 165 passengers in a single class cabin. And direct operating costs at about the same level as subsonic aircraft for intercontinental flights at the time. The potential global market was seen then as being 375 supersonic airliners, of which the US SST was supposed to sell 210 to 250 aircraft, 90 of which would be overseas, outside of America. Development

costs were estimated at being US$770 million in 1963 dollars (equalling US$6.4 billion in today's worth) for a Mach 2.2 SST and at US$890 million (US$7.5 billion today) for a Mach 3 model. Until first flight it would take six years and eight months, the US government estimated, meaning first flight was expected in 1970. The plan was then to cap government financing at US$750 million and

to recoup the development aid later with a charge of 0.5 to 1% of each supersonic ticket sold. In October 1963, Pan Am and TWA made down-payments of US$2.1 million in total for options on 21 SSTs overall.

After Kennedy's decision in favour of the SST program, it was hard for Halaby to contain his enthusiasm. "It's adding to the nation's prestige, writing another brilliant chapter of US

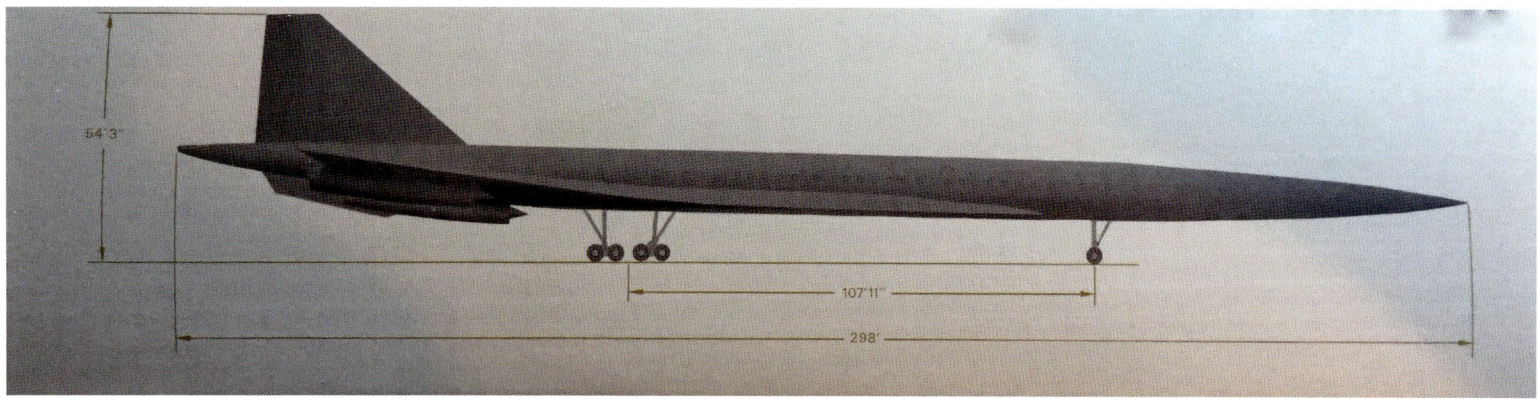

Renderings of the Boeing 2707-100 showed the gigantic dimensions of an aircraft initially 90m/298ft long, more than all of today's airliners (AS)

The 2707-100 of late 1966 was the most radical SST design ever with swing wings (AS)

aviation essential to national interest. It is a challenge to the free-enterprise American industry to show it can compete and beat nationalistic efforts of France and the United Kingdom and, perhaps, the Soviet Union," Halaby stated jubilantly in June 1963. Shortly before his assassination, Kennedy appointed two financial experts to oversee the SST budget and the FAA officially opened an office for SST development. Victory in the first dispute went to the industry – as its share of development costs was lowered from 25 to ten per cent. In November, around the time of Kennedy's assassination, Congress released US$60 million for the development of SST prototypes, while the FAA started a tender. It stated the specifications, which once again had been slightly modified: speed Mach 2.7, range 5000 km and a capacity for 200 to 250 passengers. Three aircraft manufacturers took part in the tender: Boeing, Lockheed and North American, also three engine suppliers; General Electric, Pratt & Whitney and Curtiss Wright.

On December 19th, 1963, the financial experts appointed by Kennedy, now reporting to his successor President Lyndon B. Johnson, warned: "The USA should not get engaged in a race with the manufacturers of Concorde to produce the world's first SST." In January 1964, the three contenders revealed their design proposals, and now it showed that it was extremely difficult to put the big promises given into reality. All designs the manufacturers came up with were disappointing, they were incapable of fulfilling the tasks set by such a margin that fundamental doubts arose about the viability of the whole undertaking. And the competition soon shrunk to just two applicants, as North American had to file for bankruptcy in May 1964 and exited the race. Its proposal, the NAC-60, had been the smallest with a length of just 59 metres. It was in principle an enlarged version of the B-70 Valkyrie bomber, including the same extendable canard wings on both sides of the cockpit to enhance lift for more stability in slow phases of flight. The cabin could have accommodated 187 passengers, the maximum take-off weight was set at 217 tons, the cruise speed at Mach 2.65 (2820 km/h).

That left Boeing from Seattle/ Washington State and Lockheed from Burbank/California in the race. Already by spring 1964, the commission consisting of 210 experts tasked with choosing the winner had let slip its preference for the design of Boeing and General Electric. Lockheed was already experienced with supersonic aircraft as it had introduced the F-104 Starfighter, one of the

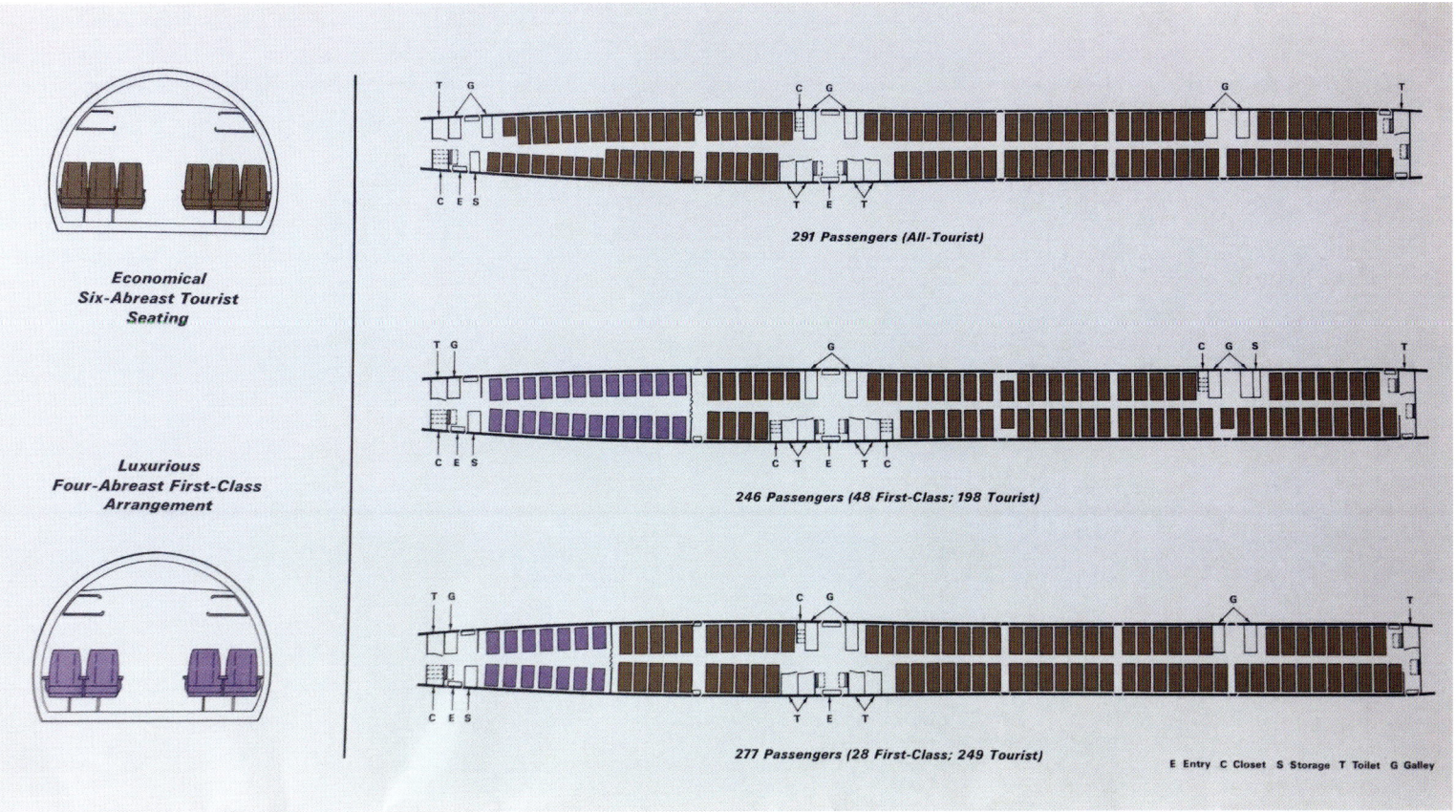

Economical
Six-Abreast Tourist
Seating

Luxurious
Four-Abreast First-Class
Arrangement

291 Passengers (All-Tourist)

246 Passengers (48 First-Class; 198 Tourist)

277 Passengers (28 First-Class; 249 Tourist)

E Entry C Closet S Storage T Toilet G Galley

A 1966 Boeing brochure shows possible cabin configurations, offering space for up to 291 passengers in the 2707-100 (AS)

first military planes flying faster than the speed of sound. Back in 1956, Lockheed engineers had conceived studies for supersonic airliners, initially with stub wings as in the Starfighter, enhanced by canards, which proved to be aerodynamically challenging. In the early 1960s, this research led to the concept CL-823. This then in 1964 became the enlarged and improved version L-2000. This aircraft came across as fairly conventional and resembling Concorde in a way. Only it was

much bigger with a length of 70 metres in its first draft, which was stretched to 83 metres in the final versions L-2000-7A and -7B; this would have allowed for up to 271 passenger seats on board and would have been longer than all aircraft existing today. The L-2000 had a double delta wing with an area of 875 square metres and a sweep of 65° to 84°. Lockheed had teamed up with Pratt & Whitney as engine supplier and Pratt had developed the new afterburner turbofan JTF-17A, which was

producing much stronger thrust and would allow for a cruise speed of Mach 3. Lockheed favoured this engine type, but it was uncertain if competing supplier General Electric could win the race with its equally new GE4 afterburner turbojet. So Lockheed decided to install four long engine nacelles under the wing's aft trailing edges, which could have incorporated both engine types. Model L-2000-7B was given another stretch to 89.3 metres, while the maximum take-off weight remained at

267 tons. The relation of lift and drag, the so-called glide ratio, indicator of aerodynamic excellence, increased to 8:1. The slightly ovoid fuselage cross-section measured 4.75 metres of cabin diameter at the widest and 3.45 metres at the narrowest point. The seating layout varied, according to cabin class and location in the cabin, between 2-2 and 2-3 per row.

In May 1964, President Johnson assigned Boeing (more on that soon) and Lockheed

In mid-1966, Lockheed unveiled the L-2000 mock-up in Burbank in front of the enthusiastic work forece (GT)

This early cutaway-model of the Boeing 2707-100 shows the extremely heavy and complex hinge mechanism of the swing-wing removed here (GT)

to come up with more detailed plans of their respective designs, ushering in phase 2A of the government's SST program, for which US$100 million were earmarked. As still no clear decision had been taken for one of the competing concepts, Johnson launched the 18-month detailed design phase on July 1st, 1965, assigning another US$200 million. Now in-depth research was conducted as well as wind tunnel tests, and most importantly the building of mock-ups. These original-size models were the only way for engineers to examine the technical feasibility of a concept before the age of 3D computing, which can simulate and test all kinds of configurations almost effortlessly in comparison. Mock-ups were also important tools for marketing departments to test a concept's customer appeal and to drum up commercial interest by potential buyers. The mock-ups of the SST contenders were setting standards. As the encyclopaedia *Jane's All the World's Aircraft*, the authority on anything flying, was noting in its 1966 edition: "A full-scale mock-up of the Model 2000, completed in mid-1966, is claimed to be the most complete engineering mock-up of a future airliner ever built. The interior is fitted with first-class and tourist-class passenger furnishings, service facilities and a replica of the flight deck, with simulated controls and

instruments. The nose can be drooped and the landing gear is retractable." The L-2000 mock-up, mostly built of wood, was absolutely massive and awe-inspiring, as can be seen in vintage promotional film clips still available on YouTube, even though it was built with just one wing due to space constraints. It is most impressive to watch the gear retracting and how tiny humans appear beneath it. Also the extreme height of the landing gear is remarkable, which would have been necessary to avoid tail strikes on take-off and landing due to the high angle of attack. The airframe of the L-2000 was positioned about twice as high above the tarmac as the Boeing 747s. Even today, the L-2000 would still have been a giant, turning heads and garnering attention, not only due to the fact that on average 225 passengers could have covered 6500 km on it in just about 150 minutes. "Lockheed's double delta-wing mock-up was similar to the Concorde configuration, but altogether much bigger. It was a massive monster," recalled Concorde test pilot Brian Trubshaw in his memoirs.

But then, the L-2000 never got further than an impressive mock-up. As on December 31st, 1966, after a further two months of examinations, by now involving 240 experts, the FAA announced its decision: the winner of the competition was Boeing's 2707-100 design with General Electric engines. The reasons: The Lockheed design would have been easier to produce and less risky overall, but its operational performance at take-off and cruise altitude were slightly inferior to the Boeing design. Because of its JTF-17A engines by Pratt & Whitney, the L-2000 was also deemed to be a bit noisier. Additionally, the FAA viewed the Boeing concept as more innovative and therefore closer to the core assignment, which was primarily to build an aircraft superior to Concorde. It

could be, however, that the DNA of the L-2000 might reappear over half a century later: from 2022 a test aircraft called Lockheed Martin X-59 QueSST is supposed to fly as part of a NASA research program and demonstrate that supersonic flight is possible with minimal boom impact as well (see Chapter 11).

Brian Trubshaw judged in hindsight: "The Boeing 2707-300 was not very different from its Lockheed competitor." But it was still a very long way to the finished design of

the 2707-300, as the winning Boeing SST 2707-100 concept, presented in late 1966, was all but ready to go into production and still had years of setbacks and modifications before it. Only now was the real work beginning, and it was one of the biggest efforts in the history of the American aerospace industry. And that at the same time as developing the Boeing 747, whose role as the primary passenger transport the SST would soon take over, according to the assumption then. Not to

forget, also at the same time, the peak of the Apollo program for a moon landing was coming up. "We were going to the moon … and there was just this whole belief in America that there was nothing we couldn't do, that whatever we set ourselves to, no matter how much it cost or how much work it was, there were no limits," says Boeing historian Michael Lombardy today. "The SST is a symbol of that period when we could do anything we were planning on doing, and it was the biggest program at

Boeing. Other than Apollo, [the SST] was the biggest aerospace program in America," according to Lombardi. "We wanted to go to the moon and built the 747, but the 2707 was always the number one project at Boeing. Joe Sutter, the 'father of the 747', frequently complained about how difficult It was to get engineers to construct his aircraft, as everyone was always busy with the SST."

Boeing's original SST design was the most radical supersonic concept ever –

In September 1966 the mostly aluminium mock-up of the Boeing 2707-100 was finished in a hangar on East Marginal Way in Seattle, close to Boeing Field (GT)eine voll eingerichtete Kabine. [GT]

because it had swing-wings. This idea came from military supersonic development and was necessary due to a simple fact: the delta wing, most suitable for supersonic cruise, does not serve slow phases of flights well, especially on takeoff and landing. So, the solution seemed to be a variable wing geometry, adjusted according to flight phase, to achieve the best aerodynamic characteristics at each point. Boeing first had developed a swing-wing design in 1959 during the military TFX competition that was won by General Electric with the F-111. In 1960, Boeing held an internal tender and the existing variable wing design was selected for a future SST, the draft was named Boeing 733 but it became publicly known as Boeing 2707. The aircraft strongly resembled the B-1 swing-wing bomber under development then and was finally introduced in 1986, with the difference that the engines were mounted in individual nacelles and not in pairs as on the B-1. The base model 2707-100 was laid out for a takeoff weight of 307 tons, range with 235 passengers was supposed to be 5500 km, the list price US$35 million per aircraft. With its length of 93.27 metres, the 2707-100 would have to this

day been the longest aircraft ever built; the current record holder Boeing 747-8 measures just 76 metres. In contrast to the Boeing 733, the fully retracted swing wings morphed into one surface with the aircraft's tail area of vertical and horizontal stabiliser, building a continuous delta wing for cruise. In low-speed flight with fully extended swing-wings and a sweep of 20°, wingspan would have been 55 metres, while in cruise with retracted wings and a sweep of 72° it only came to 32 metres.

The four engine nacelles were now mounted as far aft as possible under the trailing edge of the wings, after earlier configurations had put them further forward, potentially posing problems with cabin noise and the hot exhausts in the tail area. However, positioning the four massive engines in the very aft underneath the stub wings caused the aircraft to become tail-heavy, which had to be overcome through construction changes. The four turbojet afterburning General Electric GE4/J5P engines, each delivering 281 kN in thrust, were quite similar to the propulsion of the XB-70 Valkyrie. In civil use, the afterburners would only be ignited briefly for takeoff and to pass the sound barrier, in contrast to the bomber. Thrust was sufficient even without them for a cruise speed of Mach 2.7. The engines were designed to function reliably under extreme conditions, which civil propulsion had never been exposed to before. 70% of its running time was supposed to be at supersonic speeds, the highest stress level for the engines. As the US SST was designed for much higher speeds than Concorde which would only reach Mach 2.2, all US designs needed to use other materials. Steel alloys or titanium were the only materials able to withstand extreme temperatures of between 200°C and 250°C on the airframe, created by the friction heat at Mach 3, without loss of stability. One square metre of titanium aircraft skin would have weighed 2.2 kg, steel of the same size a full seven kilos, three times as much. So the decision was made in favour of titanium, despite it being more difficult to process and considerably more expensive. But it was well suited for the estimated 50,000 hour-lifecycle of the airframe in being corrosion-proof and crack-resistant.

The original fuselage cross-section of the Boeing 733 was now widened to a maximum of 5.08 metres. This created the first wide-body aircraft with two aisles, making it possible to install a layout of 2-3-2 per row for most of the 55-metre long cabin. At narrower points, 2-2 seats for First Class or 3-3 for Economy would have been provided, and even there, the cabin was still 1.22 metres wider than that of the Boeing 707. For the first time, there was provision of an inflight entertainment system with retractable TV screens above every sixth seat row in Economy, while in the 30-seat First Class every pair of seats would have featured an individual screen in the middle console. A precursor of in-seat-screens that only started to appear two decades later. On grounds of structural integrity the windows were tiny and only measured 15 cm in diameter, while the inner window frame was double that size to create an illusion of bigger windows. All these features could be inspected already in September 1966 in a giant aluminium mock-up in a hangar at East Marginal Way in Seattle, close to Boeing Field. Its cabin was equipped with 277 seats. "The Monster", as the oversize model was dubbed, quickly became a local attraction. Boeing often offered the general public tours, people then queued up patiently in long lines in front of the huge hangar doors. These days, the only preserved segment of the mock-up, the 26-metre front section with the droop nose, can be seen at the restoration centre of the Museum of Flight at Paine Field in Everett, north

The gigantic mock-up of the Boeing 2707-100 had movable swing-wings. In this multiple exposure shot all three wing positions are visible (GT)

of Seattle, after a long odyssey criss-crossing the country.

The atmosphere on board the 2707 would have been comfortable, up until an altitude of 10,500 metres the cabin pressure was supposed to be the same as ground pressure. Above that, the cabin atmosphere would have equalled 1770 metres above sea level, about the same as modern aircraft with carbon-fibre fuselages offer today, the Boeing 787 or Airbus A350 for example. One of the most impressive features of the 2707 was its long droop nose, consisting of two segments and

two joints. While the main part of the nose below the cockpit was to be lowered by up to 22° to enhance pilot ground vision, and offered two windshield windows for forward vision in retracted form, the forward and smaller part had its own hinge. This kept this nose segment at a constant angle towards the main fuselage, to ensure functioning of the weather radar, which was installed here. It also guaranteed sufficient ground clearance for the long nose, which was hovering 2.67 metres above ground in retracted and 1.22 metres in lowered position.

The pilots were supposed to be sitting 4.27 metres above ground, a camera under the fuselage was planned to transmit live footage of the ground situation and the state of the main gear to the pilot's workplace.

However, it quickly emerged that the concept was burdened with an enormous weight problem. Fully loaded in regular operations, it would have just had a range limit of below 5000 km and thus be incapable of crossing the Atlantic. As one of the main culprits for the overweight, the complex and

heavy hinge mechanism of the swing wings was identified. It burdened the 2707-100 with about 20 tons of extra weight, when the empty weight was 130 tons, enabling just 34 tons of payload. Three hydraulic actuators would drive the mechanism to alter the wing configuration, extending them to maximum span taking two minutes, swinging them backwards one minute, in the worst case, one actuator should achieve that on its own in three times as long. A warning system would have stopped movement of the wings from a certain critical limit, if an asymmetry had been detected. Still, there were doubts after some serious accidents with the F-111 fighter and its swing-wing mechanism if it would have been up to daily flight operations, and how much extra maintenance effort would have been needed.

In February 1967, the FAA had already requested interested US airlines to support the early phase of prototype construction for the SST by contributing risk capital of US$1 million per delivery position booked. Ten airlines, holding a total of 52 options, went along, altogether Boeing counted 113 delivery slots allocated to 26 airlines. At the same time, the FAA, Boeing and General Electric had signed contracts to build two identical swing-wing prototypes. By the FAA's next design review in November 1967, Boeing

had substantially revised the design and was now revealing the 2707-200, a giant stretched again by 3.50 metres to a length of 96.92 metres, with a planned take-off weight of 340 tons and accommodating up to 350 passengers. The main driver for the modifications was the quest to increase productivity with higher payload and less weight, enabling a transatlantic range of 6840 km. Besides lengthening the fuselage, the new version was equipped with two extendable canard wings to make up for the missing elevators on the tail. Massive weight problems remained, however, as realistic calculations of everyday operations made a weight of 385 tons more likely. And there was no solution for the mechanical and weight problems of the swing-wing mechanism. Initially, this precise variable wing geometry had been the cutting edge of Boeing's design versus the Lockheed study L-2000, as it seemed to promise higher performance and range. Now it showed that already the integration of the complex swing mechanism was posing insurmountable problems. For the highest efficiency of the extended wingspan, the hinges of the swing-wing have to be as close to the middle axis of the aircraft as possible. But this was not compatible with the construction plans, as it conflicted with wheel wells and the engine positioning. It

Not only humans, but also a replica of the first Boeing Model 1 aircraft with 16 meters wingspan bring it home huge colossal the Boeing 2707-100 would have been (GT)

became clear that the exorbitant weight of the swing mechanism wiped out all the advantages the variable wing geometry was supposed to deliver. The result of an analysis by a team of FAA technicians in January 1968 was devastating: increasing the weight of the 2707-200 had such a negative impact on the payload that the range would have been below 4500 km fully loaded, meaning the aircraft would not be able to cross the Atlantic. And therefore, the design was far below the requirements set as prerequisites for launching phase III of the US SST program, which called for construction of two prototypes. Boeing was given another year to come up with an improved design. "Boeing's Dash 200 was an exciting, challenging design, a bid to translate into reality (all) sorts of advanced concepts," wrote Kenneth Owen. "It was at the leading edge of technology. And it proved to be very difficult indeed."

It was clear that any practical way to use adaptable swing-wings still needed many years of technological progress, and that the only way to get an SST in the air in the medium-term future was to drop them altogether. In October 1968, Boeing announced therefore that it would equip its 2707-300 with a conventional delta wing, very similar to that of the Lockheed L-2000. Measuring 96.01 metres, the -300 came

out almost a metre shorter than before, its empty weight was set at 290 tons and maximum payload at 31 tons. Up to 292 passengers, 28 of them in First Class, were to be transported with Mach 2.7. The delta wing with an area of 716 square metres had rounded edges and a leading edge meeting the fuselage with a sweep of 50.5°. The wing's leading edges were equipped with slats to improve lift, while the trailing edges would have had flaps to provide high lift at slow speeds and additional "flaperons" for roll control. The wingspan (38.70 metres) was almost as wide as the Boeing 707's (39.88 metres).

Interestingly, the fuselage featured a continuous width of an identical 3.45 metres now, while the fuselage height decreased from front to back due to aerodynamic reasons. The main gear was now to have only two instead of four legs, due to the lighter airframe, and the design featured a conventional tail section. The four GE4 engines, each about 8.50 metres in length, were centrally mounted underneath the wings. They had otherwise remained unaltered throughout the program, at the time it would have been the most powerful engine ever built, delivering static thrust of 222 kN without and 290 kN with ignited afterburners. The engines were immensely fuel guzzling, burning 14,200 litres per hour without afterburners and an incredible

The Boeing 2707-100 mock-up was dubbed "The Monster" and became a massive visitor attraction, with groups touring the hangar (GT)

35,000 litres with afterburners ignited. In comparison, the new GE9X engines of which there are two on the modern Boeing 777-9, are delivering 454 kN thrust each. Total propulsion performance of the 777-9 is not hugely different from that of the 2707-300 with four engines. Only that today's engines just

consume half the kerosene per hour for double the output, compared to half a century earlier. Until 1971, the six GE4 engines built thus far ran for about 500 hours of testing, the maximum achieved thrust was 316 kN. The air inlets of the engines required a complex ramp mechanism to regulate

the amount of air streaming into or bypassing the turbines, according to the amount of air required in different phases of flight. At slow speeds, the inlets were fully opened, at supersonic speeds only partially.

A commission of a hundred experts from the FAA, NASA and the Pentagon decided

Exactly like on the L-2000 mock-up in Burbank the Braniff flight attendants also posed on its 2707-competitor at Boeing in Seattle (GT)

in April 1969 that Boeing's latest design now fulfilled the requirements for phase III. In September 1969, building yet another original size mock-up started, now of the 2707-300, as well as preparations to build two prototypes, two years behind schedule. "The Dash 300 appeared to be an economically viable aircraft, though the expected economic margin had shrunk," wrote Kenneth Owen. The first flight of the American SST was now planned for March 31st, 1972 at the latest. But in the meantime, the mood in America towards the government investing huge sums to get an SST in the air began to change dramatically, compared to the euphoric early days during the Kennedy era. "Subsidizing such a program was of course foreign to a country that relied on free enterprise control, whereas Europe accepted the government's presence," explained Brian Trubshaw. Also the airlines themselves displayed a lack of endorsement, he felt: "Few if any of the major airlines were particularly enthusiastic about being pushed into the supersonic era when they were still paying for their recently acquired subsonic jets."

In January 1969, new US President Richard Nixon had come into office and it was his decision whether more state money was to be pumped into

With the final iteration of the Boeing 2707-100 being ab ut 90 meters long, photographers needed a wide-angle lens and a look from above (GT)

the SST program. At the time he took office, shortly before Concorde's first flight in March 1969, the assumption was that the SST would take to the air in 1973 for the first time and would see entry into commercial service in 1978, whereas Concorde was stated to be entering airline service in 1974. Until 1990 the hope was then to deliver 500 SSTs. Total cost

Compared to the notoriously cramped Concorde cockpit, the pilot's "office" in the 2707-100 was outright spacious. It was supposed to be operated by a three-man cre, in the foreground the flight engineer's panel (GT)

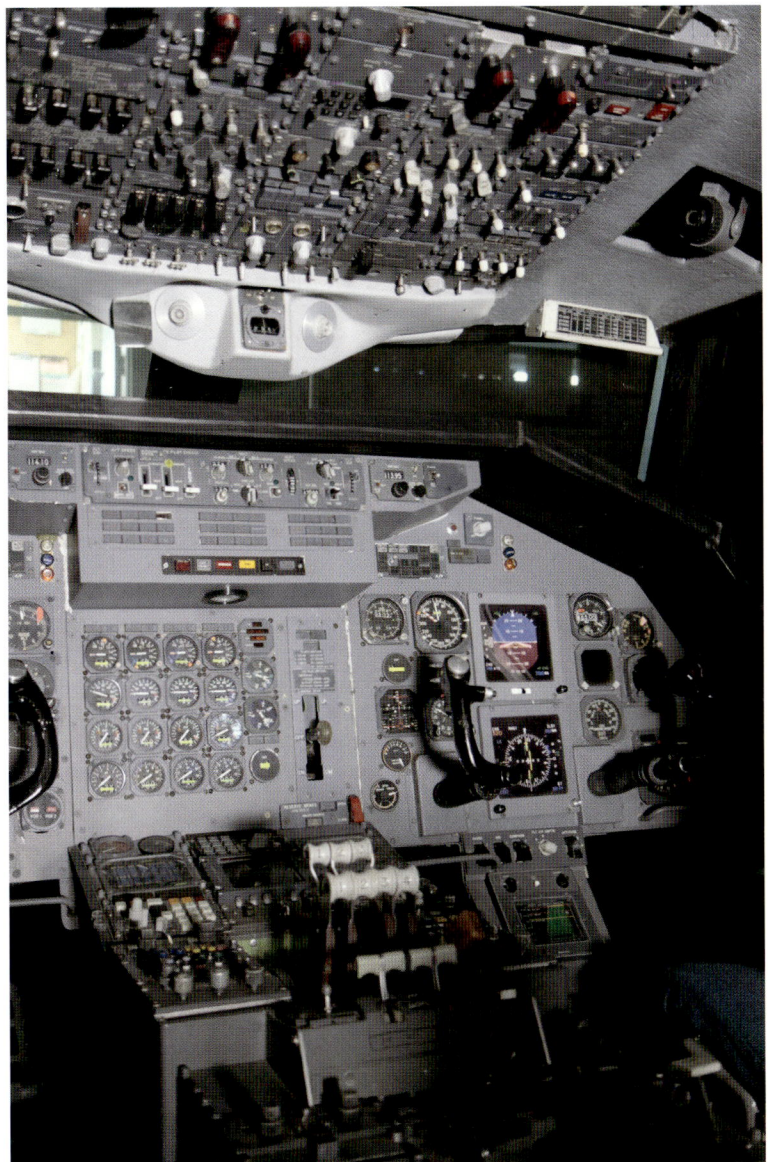

For a mid-1960s design the 2707-100 cockpit appeared very orderly. Clearly visible the white throttles for the four engines (GT)

for both prototypes alone was estimated at US$1.7 billion. A further US$4 billion, initially 89% pre-financed through taxpayers' money, would have been needed before entry into commercial flying. For the lengthy SST flight test program both prototypes and five production aircraft were earmarked, which altogether were expected to operate 4000 flight hours to become certified for passenger service. The Boeing 747, in contrast, only needed about 1400 hours until it received its type certificate. At the time there were 78 options for the European supersonic airliner and 112 for the American SST, almost unchanged since 1967, of which 58 came from abroad. Until June 1970 the count was for 122 SST options from 26 airlines. The biggest prospective customers were Pan Am (15) and TWA (12), even Air France and BOAC (long-haul predecessor of British Airways) each secured six, Lufthansa three. The price charged per

Compared to Concorde, the cabin of the Boeing 2707 was supposed to offer almost wide-body comfort (GT)

For the first time, the Boeing 2707 was supposed to offer built-in inflight entertainment with a TV screen above every fourth seat row (GT)

aircraft then was US$48 million (about US$322 million in today's worth), while the Boeing 747, also built in Seattle at the same time, was offered for US$25 million.

Soon, dark clouds were gathering above the SST program, which then strengthened to become the perfect storm. "Airline interest in supersonics, always lukewarm, waned as Boeing began to market the huge 747," wrote Kenneth Owen. Besides all economical and technical doubts, a political debate was catching momentum, fuelled by a so far unknown motivation: environmental protection. "An unprecedented campaign of opposition to new technology on environmental grounds engulfed what fragile support there was," according to Owen. And the arguments of the newly formed environmentalists were manifold and hard to dismiss: the decisive problem was noise, both close to the airports, but mostly caused by the sonic boom. "That would have been substantial for a giant like the 2707, as it depends on mass. And the Boeing design was massive, equalling today's Boeing 777 in takeoff weight," says Bernd Liebhardt, supersonic specialist at the German Aerospace Centre DLR. Other concerns of the environmentalists touched topics that are still heatedly discussed these days: a threat to the ozone layer in the upper atmosphere through emissions at altitude, as well as an increased greenhouse effect caused by contrails. And finally, they questioned alleged dangers to crews and passengers of supersonic airliners due to cosmic rays.

President Nixon instructed a committee made up of representatives of different government departments, led by the secretary of transportation, in April 1969 to review the SST program. Their verdicts gave a mixed message, but on September 23rd, 1969, Nixon announced: "I have made the decision that we shall go ahead. I have made it because I

This interesting scale comparison of the final Boeing 2707-300 without swing-wings (in the back) as well as the Tupolev Tu-144 (middle) and Concorde (front) shows that the Americans intended to play in another league (GT)

America's premier airline Pan Am had initially signed options for Concorde, then also for the Boeing 2707, to which it dedicated its own model (AS)

make any sense?" Proxmire asked rhetorically. His Senate colleague Gaylord Nelson summed it up as follows: "This is an expensive aircraft with high fares for the small clientele that is prepared to pay a high premium to save three hours of travel time to Europe. If there is really enough demand for such an aircraft, it should be built without subsidies." In a congressional hearing chaired by Proxmire in May 1970, an IBM physicist announced: "The SST will create noise equalling 50 Jumbo Jets taking off." The assigned SST budget for 1971 was passed in Congress in May 1970 with a majority of just 13 votes, but in early 1971, the SST project had lost the support of both the public and politics. Even America's greatest pilot hero, Charles Lindbergh, first to cross the Atlantic non-stop in 1927, was now taking sides with the environmentalists.

In polls 85% of respondents voted against it, while not even half of all Americans had ever flown. At that time, already over one billion dollars had been pumped into the SST project, of which US$150 million came from Boeing itself, and 8.5 million man hours had been invested. For the first parts of the prototypes, metal had already been cut and about 15% of the airframe built. Around 1500 Boeing personnel in Seattle were dedicated to the SST at the time, which, including suppliers, gave jobs

want the US to continue to lead the world in air transport." For 1970, Congress was expected to appropriate US$96 million and a further US$662 million up until 1974. And the FAA was supposed to come up with a total of US$1.4 billion for construction and testing of two prototypes, which would equal the unimaginable sum of about US$35 billion today. And soon the dispute was in the middle of the floor of the US Senate. One of the leaders against the SST program was Senator William Proxmire. "We are supposed to spend 290 million dollars for 1971 alone so that a half dozen people, the jetsetters, can fly overseas, while we spend 204 million dollars for urban public transport, taking millions of people to work, does that

to an estimated 14,000 people, the reason why unions were supporting it up until the very last. In early 1971, a bitter battle was fought between lobby organizations of both sides, at the same time the House of Representatives introduced a new rule stating that members weren't able to vote secretly any more. This probably was the last part that turned the balance, when on March 18th, 1971, the vote came out about the SST program: the opponents won with 215 to 204 votes, if by a surprisingly narrow margin. Six days later the Senate voted 51 to 46 against the supersonic airliner as well. These votes did not outlaw the building of an SST, but Boeing would have had to finance it with private capital. But as the manufacturer had to finance the simultaneous development of the 747 the same way, the company was mortgaged up to the neck and

In September 1969 construction began on a new mock-up of the final Boeing 2707-300 design, now featuring delta instead of swing-wings and 96 meters in length (GT)

The colours of the Braniff cabin crewmembers in the mid 1960s miraculously matched the colour palette of the seat covers in the Boeing 2707 (GT)

had no room whatsoever for financial manoeuvring. The American dream of a pole position in the anticipated supersonic age had burst like a bubble.

In summary, the SST adventure is sobering. The Boeing 2707 wanted too much too early, at a time when many of the necessary technologies where only emerging. "When we built Concorde," said Kit Mitchell, who was part of it, "we pushed technology as far as possible at the time. Boeing, to the contrary, pushed something that was simply too difficult." Kenneth Owen stated looking back: "The abortive US SST may well be the most discussed, most thoroughly analysed aircraft never built. It had cost American taxpayers more than US$1.35 billion. For about the same total, Britain had shared the complete development of the Concorde up to the granting of a full Certificate of Airworthiness." Also, Concorde test pilot Brian Trubshaw stressed: "What few people realise is that the USA spent more money and finished up with nothing than was ever spent on Concorde." Even the initially glowing SST advocate Najeeb Halaby had to admit: "An American SST would have been a financial disaster for all concerned: Manufacturers, governments and airlines alike." Former Boeing CEO Phil Condit was openly admitting in 1996, 25 years after: "Boeing came within an eyelash of bankruptcy." Company historian Michael Lombardi assessed: "What finished off Boeing's design and also Concorde in the end was their fuel burn. That was just unbearable. In the 1971 recession the oil price began to rise. If that wasn't the end, then the 1973 oil crisis would have done it. It would have ended in disaster if it would have been continued."

Truly disastrous was the end of the SST for many people in Seattle, the resulting "Boeing bust" is part of the local mythology until today. Including suppliers and other industries, over 60,000 jobs were lost. "Boeing was looking at cancelling the 737 because it wasn't selling. The 747 was having a hard time and so it was a pretty dark time," recalls Lombardi. An exodus from the city was the result. Locals trying to be humorous had a big billboard put up next to the major highway with a request printed on it: "Will the last person leaving Seattle – Turn out the lights." In the end, both the 737 and the 747 became huge global success stories and are still being built in 2021, while the SST saga soon fell into oblivion for most.

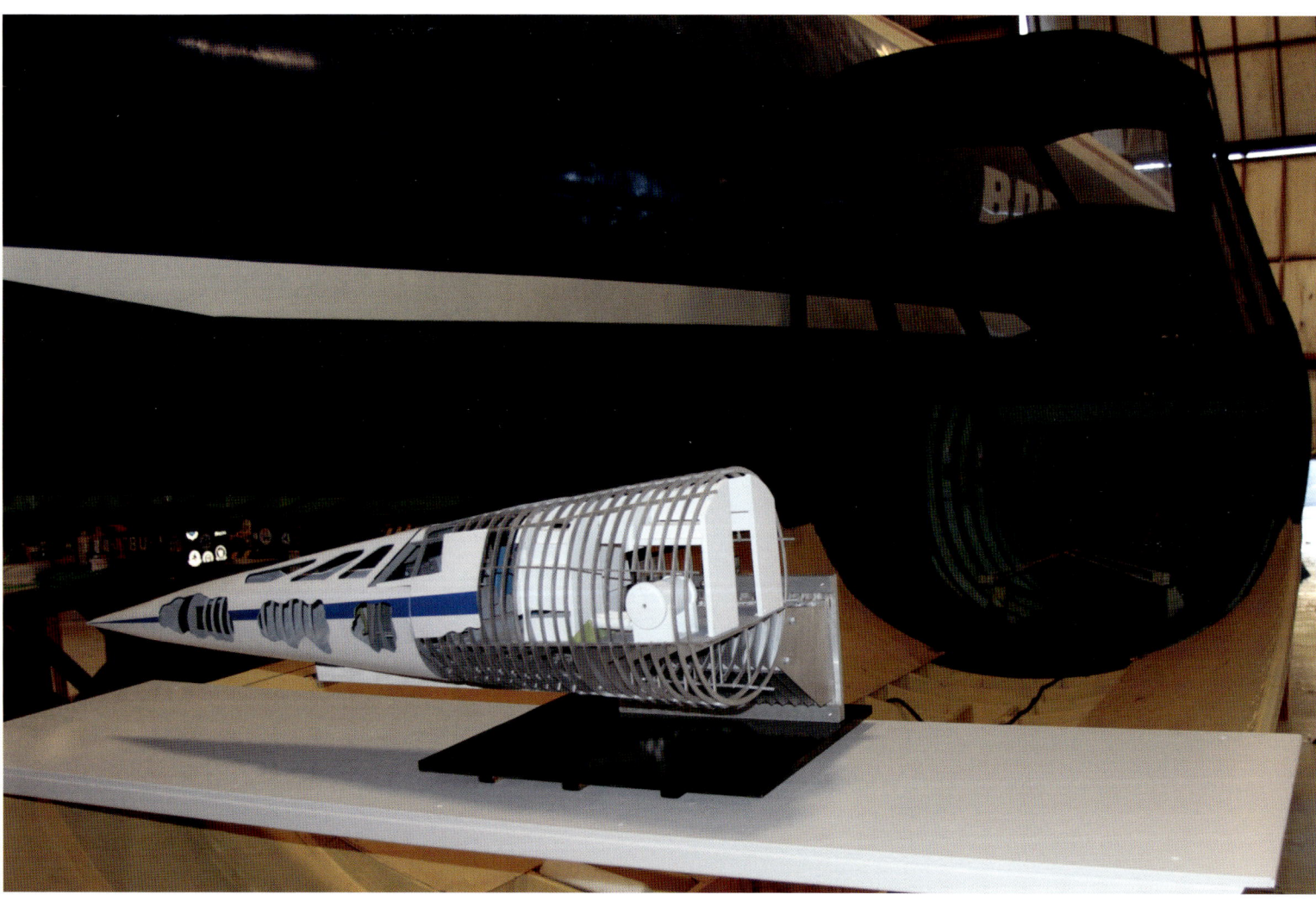

The Museum of Flight plans to reconnect the droop nose with the remaining front section of the 2707-300 mock-up. A model shows its structure (Ted Huetter/Museum of Flight)

The dimensions of just the landing gear and the front section of the 1969 mock-up of the Boeing 2707-300 give an idea of what a giant this 96-meter supersonic airliner would have been (GT)

The role of the USA – Boeing 2707 & Lockheed L-2000

four

The world's first supersonic airliner: The Tupolev Tu-144

"In terms of investment and return the Tu-144 may well rank as the biggest single failure in the whole history of aviation."

Western expert in Moscow, 1980s, quoted after Moon

The second test aircraft of the Tu-144 at the Paris Air Show in June 1973 just before it crashed ©Ad Jan Altevogt

Impressions of the Tu-144 prototype in production which began in Moscow Ramenskoye (later Zhukovsky) in 1965 (Tupolev PJSC)

To comprehend what an unimaginable effort, but at the same time total absurdity the development of the first supersonic airliner in aviation history represented, one has to put oneself back into the 1960s Soviet Union. Because it actually was the Soviets, fuelled by the Cold War and a competition between East and West, that first brought an SST into the air on December 31st, 1968. Over two months prior to Anglo-French Concorde achieving first flight on March

2nd, 1969. Exactly that, being first was the primary goal, and, as emerged during the following decade, it would remain the sole and most essential success of the Tupolev Tu-144. It was a remarkable success, hitting Europe and America unexpectedly, although foreseeable. In hindsight it appears almost insane that the Soviet Union, the way it was then, opted to develop and market a supersonic airliner. At a time when it was also struggling to keep the upper hand in the military race for nuclear arms supremacy as well as succeed in being the first to bring a man to the moon. The world's biggest country and only super-power besides the USA, was also the one with the worst roads, the fewest cars and the patchiest rail network. In winter, only very few driving lanes were even usable and, besides the main artery between Moscow and Minsk, there were few multi-lane overland routes. Only the Trans Siberian railway tied the Western end of the Communist empire with the East, travel time from Moscow to Vladivostok: one week. At the same time, there were 50 daily flights on this route. State carrier Aeroflot, operating a fleet of over 3000 aircraft, three quarters of them with piston-engines, was by far the biggest airline in the world. In 1966, it had already carried 48 million passengers, in 1967 this had risen to 53 million.

Production of the first pre-serial aircraft CCCP-77101 in Moscow about 1970 (Tupolev PJSC)

Its fares were extremely low because they were heavily subsidised. Black market dealers were able to reap a profit flying a basket of fresh cherries from Crimea to Moscow and back after selling it. Aviation was the lifeline of the huge empire, and the Soviet Union boasted a world-renowned aerospace industry and prominent design bureaus such as Mikoyan, Sukhoi, Ilyushin, Antonov and Tupolev. Legendary designer Andrey Tupolev (1888-1972) had already enabled Soviet aviation to make a quantum leap in 1956. As the world's second jet airliner (after the British *Comet*) he brought the Tupolev Tu-104 for up to 115 passengers to market, due to political reasons also in haste and before it was technologically mature. Prior to

that, Aeroflot only flew propeller aircraft with an average cruise speed of 200 km/h. In 1966, a decade after introduction of the Tu-104, on average 85% of all passengers flew in jets at about 550 km/h; by 1968 the average speed rose further with faster aircraft. Still, the journey from Moscow to the Far East remained an arduous one. To cover an air distance of over 6000 km, the Tu-104 had to make two fuel stops, so the jet trip was only marginally quicker than flying in a propeller-driven Tu-114, needing eight hours for a nonstop flight. On average, every Aeroflot passenger saved 24 hours of travel time versus the quickest alternative surface transport. That equalled a time gain of 55 million man-days or the yearly labour time of 150,000 workers. And the quicker travel times in the air created extra traffic: between 1956 and 1967, passenger and cargo volumes of Aeroflot increased nine-fold. This also meant a halving of operating costs, the income per passenger rose by 10% and the profitability of the airline doubled, although Aeroflot wasn't a profit-oriented enterprise in a Western sense. Swiftness and efficiency were usually absolutely foreign to Aeroflot, where in Soviet times endless delays, irregularities and a harsh tone dominated. Still the experience of fast jets of the first generation created an appetite for more among the Soviet elite – and the idea of a homegrown supersonic airliner gained traction.

"The number of high-level Soviet leaders associated with it attests to its appeal to the official Soviet mind-set; it represented the values and dreams of a generation of Soviet elite," wrote Howard Moon in his leading work "Soviet SST", published in 1989 even before the Iron Curtain fell. "Its speed, power, prestige and glamor were to be a showcase for rapidly advancing Soviet technology." But to demonstrate this with an SST of all things appears contradictory at first sight, as a supersonic airliner is much more expensive to operate simply due to higher fuel burn. In the West, the assumption was of a lucrative market niche of time-sensitive travellers for whom saving several hours would

All smiles after the Tu-144 first flight in Zhukovsky on December 31, 1968. Third from right is chief designer Aleksey Tupolev (Tupolev PJSC)

be worth a hefty premium. In the Soviet Union, however, the formula "time is money" was totally meaningless. Why then should its aerospace industry, already running near its limit under extreme pressures, tackle such a gargantuan task? Especially as the race to bring a man to the moon had the industry focus on long-range

missiles for high altitudes rather than supersonic bombers. This forced the manufacturer to develop the Tu-144 as a standalone civilian product, rather than adapt an existing and tried military program.

From the beginning, politics and propaganda played a decisive role fuelling design and construction of the Tu-144, to

beat "the West" in the perceived race to take the first supersonic airliner off the ground. This was the ultimate goal, all else was of lesser priority, which did not benefit the aircraft's design. "For the Soviet Union it was unimaginable then to allow the West to achieve a lead, while it would stay behind," said Tupolev designer Alexander Pukhov. "We

not only had to prevent the West from overtaking us, but we had to keep it in check and try to get ahead ourselves. This was the task given to us by Secretary Nikita Krushchev." But, as was acknowledged in the West, there was an actual need for such an aircraft, especially on the long Communist prestige runs from Moscow to Beijing or Havana.

31 декабря 1968 г. впервые в мире в небо поднялся
сверхзвуковой пассажирский самолет — советский лайнер ТУ-144!

The Soviet airliner TU-144
made the world's first supersonic flight with passengers
on board on December 31, 1968!

Aeroflot claims the first supersonic passenger flight in its Tu-144 brochure happened in 1968, but this was the first flight ever with just the crew wearing parachutes (AS)

In 1961, the then Administrator of the US aviation authority FAA, Najeeb Halaby, who spoke fluent Russian, seemed impressed when returning from Moscow. But probably more so because he wanted to speed up the American project. The Soviet SST for him promised to be "the greatest, the most efficient, most productive long-range transport in history, so they have a basic Communist communication need for it." By the initial Western unveiling of a model of the Tu-144 at the Paris Air Show in 1965 it was announced that the first flight was planned for 1968. This information was spread on purpose, to clearly state the aspiration of being first, whilst the Concorde builders did not want to accept joining a race. "Political priorities to overcome the West, no matter how, obviously played a negative role, as they favoured rushing over proper scheduling in a highly challenging and complicated field," said Ilya Grinberg, professor at New York State University in Buffalo.

"Nikita Krushchev in the Kremlin had huge respect for aircraft designer Andrey Tupolev. Krushchev was aware that it wasn't about an aircraft, but rather about a quantum leap in all areas," Tupolev engineer Alexander Pukhov recalled. And it was Andrey Tupolev who later calculated the economical viability and claimed that its case was built on the man-hours a supersonic airliner would save and that the project was aimed at the vast distances in the country. The Soviet SST was supposed to shrink flying time from Moscow to Khabarovsk in the Far East, close to Japan, from eight to three and a half hours and to bring Siberia closer to the far away capital. At the time, 85% of travellers on these routes were flying, mostly bureaucrats, planners and civil servants. The five hours saved, multiplied with the number of professional travellers, resulted in an impressive efficiency gain on paper, used to advance the SST program internally. "The Soviet SST obviously was not originally intended for the toiling masses, nor were its infrequent scheduled flights truly open to the public. It was conceived as a conveyance for the Soviet elite", wrote Howard Moon.

Advisers of President John F. Kennedy feared in 1961 that the Soviets could install a passenger pod under a supersonic bomber, to expedite its leadership to a politically charged location in no time. So that Soviet leader Nikita Krushchev could miraculously appear in Delhi, Beijing or even London or Paris to demonstrate the technological prowess of his country. The US had an almost identical plan for the modification of a B-58 bomber. However, none of this ever became reality, neither in East nor West. Instead, preliminary works on civil supersonic projects had been taking place in the Soviet Union since the late 1950s, after the country had introduced mass production of a supersonic jet for the first time in 1955: the MiG-19, flying up to Mach 1.22. In 1956, the MiG-21 followed, the first fighter able to fly Mach 2.2. Early designs for supersonic bombers such as the Tu-125 or Tu-135, of which also a civil version was mulled, were not pursued. Priorities shifted towards space, therefore the preliminary works of construction bureau Myasishchev in Moscow on civil projects were terminated. Instead, Tupolev got the assignment. It already was experienced with big airliners, besides the first Soviet passenger jet Tu-104 also the Tu-114, the world's biggest and, despite its propellers, also fastest passenger aircraft at the time. Since 1960, the year before its entry into service, the Tu-114 has held the record as the world's fastest propeller aircraft, even until today, achieving jet-like speeds of 880 km/h. With the Tu-22 bomber and the Tu-128 long-range interceptor, Tupolev had already gained initial supersonic experience.

On July 26th, 1963, work officially began on the Soviet supersonic airliner, ten days after the Council of Ministers had decided in its favour. From now on the program's name was Tupolev Tu-144. It was to be the biggest and most complex aircraft program the USSR had ever tackled. Within five years, according to plan, five prototypes were to be built, the first by 1966. A MiG-21I was used as flying simulator for wing design. The specifications called for a cruise speed of Mach 2.2, range of 4000-4500 km for 80 to 100 passengers or 6000-6500 km for 30 to 50 travellers, at a maximum takeoff weight of 120 to 130 tons. The whole project hinged on the question of whether engines could be developed to deliver the enormous thrust necessary; only turbofans with afterburners would be up to the task. After Myasishchev had to abandon work on a big SST, during which

In 1975 test aircraft 02-2 was re-registered for the Paris Air Show and CCCP-77104 became the more memorable -77144 (AA)

the prototype of an efficient turbofan without afterburner had already been conceived, it was taking a risk to assign the manufacturer Kusnezov with the Tu-144's propulsion. Engineers delivered two-shaft jet engines called NK-144 with 20 tons of thrust each, advanced from the NK-12 turbofans powering the Soviet long-haul aircraft Ilyushin

IL-62. Specific fuel consumption of the NK-144 was 1.35 to 1.45 kg per kg thrust per hour, whereas before, at Myasischev, it had been established that fuel burn could only be 1.2 kg to enable the desired range. The prototype of the "16-17" turbofan designed there delivered 18 tons of thrust and only needed 1.15 kg fuel per kg thrust per hour. The Olympus

engines of Concorde, giving 17.2 tons of thrust at takeoff with afterburners ignited, needed 1.327 kg per kg thrust per hour, including afterburners. Only in 1964 was the decision made to task the Koliesov design bureau with restarting work on an efficient turbofan without afterburner for the SST that had begun earlier for Myasischev.

Even Andrey Tupolev initially fell victim to the same illusion as all other engineers who headed different supersonic design efforts described in this book: to build an SST that could be operated at costs comparable to the usual subsonic jets at the time. But mostly he wanted to be part of the race for an SST, his favourite expression was: "The

main thing is: Get in on the act," reported people who worked with him for decades.

In its design, the Tu-144 was laid out for flying predominantly over land, meaning there was a need to fly higher than Concorde, requiring more thrust and bigger wings. The Tu-144 prototype was smaller than Concorde, almost four metres shorter at a length of 58.50 metres. The wing area, however, was much larger, measuring almost 470 square metres, while Concorde did not even reach 360 square metres. Initially, a dozen different configurations were examined, also with elevators on the tail, which were dropped as it would have created a fifth more drag. So-called canard wings for more stability in slow flight speeds were not envisaged at first, as it was assumed they would impair wing aerodynamics. Numerous wind-tunnel tests with hundreds of models were necessary at the Central Aerohydrodynamic Institute (TsAGI) in Moscow, until the decision was made in favour of a design without elevators, but low-mounted, ogival double-delta wings. Its triangular wing profile was 78° swept at the front

and 55° in the aft fuselage.

The four NK-144 afterburner turbofan engines were mounted in pairs close to each other underneath the wings, each one boasting its own rectangular air inlet with horizontal ramps and an automated control mechanism. It was supposed to supply the turbine with the ultimate amount of air in each phase of the flight, a crucial prerequisite for supersonic cruise. But already the construction process proved problematic, as the engine nacelles and inlets had been designed independently of the engines themselves. This later led to a lack of coordination between these two fundamental systems and remained a permanent weakness of the Tu-144. Initially, both outer engines were supposed to be equipped with thrust reversers, but this did not happen. Thus the Tu-144 became the last passenger airliner, apart from early serial models of the Tu-134, to be routinely equipped with braking parachutes. This was essential due to a lack of slow flying skills, leading to high landing speeds of up to 369 km/h, while Concorde landed at 287 to 296 km/h. The prototype had triaxial main gears with a total of twelve wheels. On Soviet airports with less bearing capacity, it was necessary to spread the aircraft weight over a wider area, enabling the Tu-144 to operate from remote airports, also Russian tires were less resilient. The airframe of the Tu-144 consisted of conventional aluminium alloys. Parts of the wing structure as well as the four elevons (combined elevators and ailerons) at both wing's trailing edges, a heat shield at the tail close to the engine exhausts plus the wing leading edges and rudders were made out of titanium alloys or stainless steel. Titanium was responsible for 15% of the aircraft weight,

while non-metallic materials made up 23%; the airframe was conceived for 30,000 flight hours over 15 years. Similar to Concorde, the Tu-144 boasted a droop nose that could be lowered by up to 12°. This enabled sufficient pilot vision forward and downward, even at high angles of attack between 11° and 12° on takeoff and landing, due to the delta wings.

The cockpit was laid out for four flight crew, two pilots as well as flight- and test engineers. For the initial test flights, the aircraft was even equipped with ejection seats, unique for civil aircraft. Instruments such as auto pilot, flight guidance systems and avionics providing all-weather landing capabilities were the top products of Soviet technology and symbolised breakthroughs for Russian engineering at the time. Still, its unreliable and voluminous avionics consistently remained weak spots of the Tu-144.

Since the first models of the Soviet SST appeared at the Paris Air Show, a myth was building up in the West claiming the Tu-144 was a copy of Concorde. The press was quick in creating catchy expressions like "Concordewitsch" or the long-lasting "Konkordski," before anyone had even seen the real aircraft. It was inevitable that both designs had fairly much in common at first sight, as they both were aimed at the same missions and its designers

The three-man cockpit of CCCP-77112, on display at Technik Museum Sinsheim (AS)

of course had to adhere to the same basic aerodynamic conditions. Therefore it was unsurprising that the outward appearance of both SSTs came out at least comparable. But as all this was happening in the context of ideological and technological competition between East and West with the Cold War constantly present, the myth of the "Concorde copy" thrived. In fact, there was even an official cooperation between the Soviets and the French, initiated by a rapprochement between Nikita Krushchev and French President Charles de Gaulle. The deputy director of TsAGI, Georgy Byushgens, recalled an official Soviet visit to Concorde's production facilities in Toulouse. "When we arrived at the construction bureaus, our visit was organised as such that we couldn't remain standing anywhere, and especially not spend any time at their drawing tables. I am absolutely unaware of us taking

Prototype CCCP-68001 at the ILA Hanover Air Show in April 1972. The gangway marked "München" was only on loan (Hanover Airport archive)

Shortly before the drama: A Concorde takes off at the Paris Air Show 1973. On the ground Tu-144 CCCP-77102 that later crashed during a flight display (AA)

anything away. Aerodynamics and geometry of the wings were also fundamentally different." But former French Concorde program director Pierre Gautier had different memories, shared in 2003: "There was a secret competition between Concorde and Tu-144, because the Russians were hindered by a lack of suitable material for the airframe. When a Russian delegation visited the Concorde factory, I observed that our guests were repeatedly stamping their feet, and I asked our security people to check the Russians' shoes when they were leaving. And of course, there were shavings of our material embedded underneath their crepe soles."

There was in fact undeniable proof of Soviet attempts to enhance and accelerate their SST program, and it did not come as a surprise at a time when espionage on all fields and in all directions was part of everyday life. Over 20 agents of several Eastern intelligence agencies were part of "Operation Brünnhilde", aimed at securing the Concorde plans, revealed the London *Observer* in late 1969. Whereas Howard Moon acknowledged two decades later: "The extensive Soviet industrial espionage apparatus made a major contribution to this accelerated development timetable. In the early 1960s, aerospace was the first priority of Soviet technological espionage, and SST data were among the most eagerly sought," wrote Moon. "Hundreds of agents and thousands of analysts collected and sifted through information on British, French, and American SST designs. One effort, representing a pioneer use of Aeroflot as an espionage front, saw communist sympathisers in Aerospatiale's Toulouse factory steal Concorde blueprints." Time and again, Concorde spies had been caught and arrested, once the head of the Aeroflot Paris office, another time a Swiss who had gathered information from England and France for some time in Brussels and then sent it on regularly on the train toilet of the East-West Express to East Berlin. After being detained, he immediately offered to work as a double agent and led Western intelligence services to arrest a spy of East German state security (Stasi) in Paris. In 1966, two Czechs disguised as priests had their covers blown and were detained after tracking Concorde blueprints. Brilliant plots for thrillers and spy movies still today. But one fact remained unaltered: nobody could take the propaganda victory of having been first away from the Soviets. However whatever they might have been getting from their spies, it wasn't sufficient to enable them to build a supersonic airliner that was technically capable of fulfilling

its mission. "It was too late, if one would have given us two years more time, we would have built the wing differently, but now we couldn't do that anymore," confirmed Tupolev engineer Alexander Pukhov.

Building of the prototype had already begun in 1965 in Moscow, at the same time as work on a second fuselage for static testing. But only two years later the aircraft was ready to be transferred to the Flight Research Institute's test centre at Ramenskoye airport south of Moscow, today commercial

Zhukovsky airport. The cabin was initially equipped with 98 seats, but was now modified to offer 120 seats, with 40 in the front compartment and 80 seats aft. The slightly elliptical cabin cross section enabled a cabin width that was barely sufficient to install five seats abreast in 2-3 configuration. There was no space for a cargo hold in the belly, so the cargo compartment was in the tail section and was supposed to be accessible through a hatch from the ground, baggage was to be stowed in nine

special containers. At the same time, the maximum takeoff weight was increased from 130 to 150 tons and a test flight campaign with two modified MiG-21Is had begun. They now boasted the wing shape of the Tu-144 and were able to reach beyond Mach 2.2. One aircraft was flown by Tupolev chief test pilot Eduard Yelyan, later captain of the Tu-144's first flight and named "hero of the Soviet Union." And there was a self-imposed time pressure: "When Concorde's first flight date was set on February or March 1969,

we knew that we needed to get our aircraft up until the end of 1968," recalled Tupolev chief designer Alexander Pukhov.

And so, at the end of 1968, the prototype bearing the appropriate registration CCCP-68001 (for the first Tu-144 and the year 1968) was on time and indeed ready to fly. Engine test runs, taxi testing and last ground checks took a full month, but then in December, prolonged bad weather prevented the aircraft from taking off. "From December 20th to 31st, there were endless storms at

Promotional shot of the Tu-144 cabin (AA)

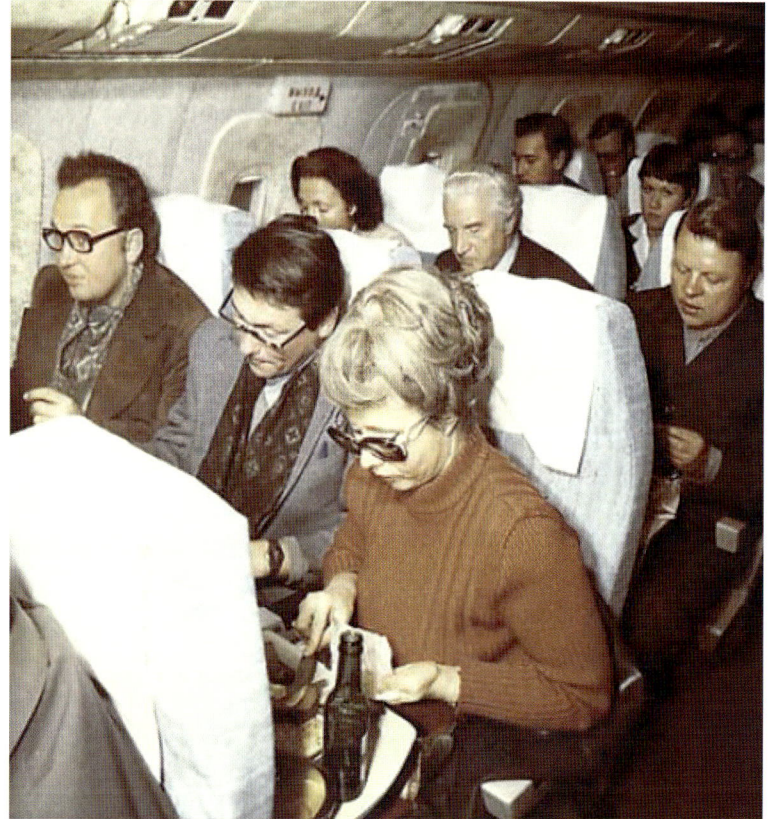

This scene from an actual flight gives an impression of the cramped passenger seating (AA)

Even though the Tu-144 still had no type certificate, supersonic proponents wanted to finally start preparations for passenger operations, seven years after the aircraft's first flight. Therefore, on December 26th, 1975, a "scheduled service' was launched on the route from Moscow to Alma-Ata, also earmarked as a passenger route. But initially just cargo and air mail were flown on the route of 3260 km, not a distance suitable for supersonic flying. And especially odd to expedite mail and cargo in a country where a few hours more didn't matter, as delivery on the ground was slow anyway. On the cargo and mail runs, speeds of up to 2200 km/h were reached at 18,000 metres altitude and initially there were two weekly flights. But the first setback came soon – a problem with fuel feed led to the hoped-for inaugural flight of passenger operations being postponed far into the future, and also the cargo flights were cut to just one weekly in June 1976. "The Tu-144 failed even in the light-duty supersonic air-freight service," stated Howard Moon. Another serious problem appeared in 1976: structural failure in static tests, where two airframes got damaged beyond repair in simulations. Back in 1971, a major crack had been detected in one airframe on returning from the Paris Air Show, while stopping over in Warsaw. It turned out that it was caused by

the production method of milling individual parts up to 19 metres long from big metal blocks made of a substandard aluminium alloy. If a crack was forming, it quickly spread to beyond one metre in length and there was no way of stopping it, as happened in 1976 during tests of a Tu-144S at 70% of the planned loads. Due to continuous problems of such kinds, the wait for the type certificate was prolonged, but finally awarded by the respective authority called USSR Gosaviaregister on October 29th, 1977, almost nine years after the first flight. A total of eleven aircraft including prototype and a pre-production models had been built so far, at least four of them were deemed suitable for scheduled passenger service.

Also in 1977, the Soviets were apparently so convinced that they were stuck in a technological dead end in developing the Tu-144, that they overcame themselves and did the unthinkable - asking the West for help. An underlying factor was the overburdening of the Tupolev design bureau, even leading to actual design mistakes. Officially, almost the whole Soviet R&D infrastructure was put to work serving the Tu-144. But in parallel, the three-engine Tu-154 airliner (having had its first flight in 1968 as well) and the Tu-22M bomber (first flight in 1969) had to be finished as well. A design mistake at the

wing roots even led to a call-back of the entire first series of Tu-154 production. A problem that the Tu-144 developers could not solve was one of the most complex for supersonic aircraft overall: their electronic, computer-regulated engine and variable air inlet controls. For these they now hoped to be able to use technology that had been developed for Concorde and had already proven itself. They requested support to improve

their air inlets from Lucas Aerospace in the UK, which had developed the engine control units, and from the Concorde consortium, consisting of British Aircraft Corporation and Aérospatiale. By 1978, the USSR wanted to order an even greater array of Concorde technology, from de-icing systems for the air inlets to water deflectors for the landing gear, highlighting their unsolved problems with the Tu-144. None of this was ever

delivered to the Soviet Union, as the British government had put in its veto, due to the well-founded concern that much of the technology could also be used in Soviet military jets such as the Tu-160.

From 1979 until 1987, Boris Bugayev was Minister of Civil Aviation in the USSR, and as the former personal pilot of Kremlin leader Leonid Brezhnev, he had a close relationship with the long-lasting head of state. Bugayev

The canard wings, extendable from above behind the cockpit like ears, were only regularly installed in a later phase of the Tu-144 development to enhance stability at slow speeds (AA)

The fully lowered droop nose of the Tu-144 was usually seen on approach and during taxi to enhance the pilot's ground vision (AA)

always had a difficult relation to Andrey and Aleksey Tupolev, however, and had never been supportive of the Tu-144 project. Instead, Bugayev's focus had been on cheap mass transport early on, and he supported developing an efficient wide-body long-haul airliner in the late 1970s, the Ilyushin IL-86. In June 1977, Leonid Brezhnev visited the Paris Air Show in Le Bourget, where this year Concorde was present, but not the Tu-144. Tupolev engineer Vladimir Vul recalled: "The French overwhelmingly praised Concorde in talking to Brezhnev, and he didn't appreciate it at all, it hurt his pride. He said to Bugayev: 'We also have such an aircraft. The French show theirs and we don't. Make sure they also do work a bit quicker so that we also can show it off and be proud of the Russian technical genius.'"

The Soviet leadership decided, due to political reasons, to put the Tu-144 into scheduled service, although it knew about the test results stating it was structurally unsafe and not suitable for regular operations. It attests to the limited significance Aeroflot was attributing to its perceived flag ship that it hadn't even been mentioned in the five-year plan for 1976 to 1980. It was not the airline leadership's decision to start passenger flying despite all recommendations to the contrary and Aeroflot followed the political orders

only reluctantly. The big day was picked as being November 1st, 1977, coinciding with the 60th anniversary of the October revolution. At 8.30am, departure was scheduled for the Tu-144's first scheduled flight from Moscow's Domodedovo airport to Alma-Ata, flight number SU499. The Tu-144 was not an easy fit for practical flight operations, it "required special passenger ramps because of its tall landing gear, and baggage had to be carried through the long, cramped passenger compartment to the tail," reported Howard Moon. Here the small cargo compartment was located, with a volume of just 20 cubic metres. And for as many years as the high tech aircraft might have been tested – the devil was in the details on the ground. After a demonstration the day before, the battery of the specially developed gangway was dead. It would not move an inch and finally, after several desperate attempts, the gangway had to be removed with raw force by a bulldozer. Minister Bugayev was furious and cancelled the banquet planned for the arrival in Alma-Ata. It was remarkable that even the commercial pilots wore special pressure-compensating vests and leather helmets enclosing their heads, while their colleagues on Concorde sat in the cockpit just in their shirts, even at Mach 2. "In a loss of cabin

pressure pilots wouldn't have lost consciousness with these vests," Tupolev engineers later pointed out. "With its four jet-engine afterburners shooting orange tongues of flames like rockets, the 197-foot-long plane lifted sharply off the runway here at 9:02 A. M., after a half-hour delay caused by a new motorized embarkation ramp that would not work," The New York Times reported the next day. "The Minister of Civil Aviation, Boris P. Bugayev, was furious."

There were 80 passengers on board, despite cabin seat maps published earlier showing a total of 138 seats in the regular Aeroflot configuration. Only two of them had paid the regular fare of 83 Russian Rubles (about $400 USD today), including a supersonic surcharge of 21 Rubles. This fare equalled about half of an average Soviet monthly income, meaning the flights were unavailable for the wider population. "Propaganda imperatives took precedence over realistic consideration of its potential economical contribution or potential costs," asserted Howard Moon. "Its role as an airliner serving the Soviet economy was eclipsed by the drive to play up a showcase of Soviet technology on a global scale." Not by accident, almost all participants in the first flight were invited official guests and members of the media, while 58 seats remained empty, as

The engine air inlets with adjustable ramps, compressing the incoming air stream depending on the situation, are the most complex parts of any supersonic airliner. The tu-144 designers were less successful on this than their counterparts on Concorde (AS)

would be the case on all future passenger flights due to weight reasons. In the front, according to the cabin seat map, the aircraft was fitted with a First Class cabin in a 2-1-layout per row with a generous 40" (102cm) seat pitch, even though First Class was never offered on domestic flights.

Most of the economy cabin in two compartments had seats installed in a 3-2-configuration per row, while in the back, where the fuselage was tightened for aerodynamic reasons, there were only two and two seats in a row for the last six rows. A few international journalists, hand-picked by the Soviet foreign ministry, among them two Americans, made use of the first, and as it turned out also last opportunity for any Westerner to

test the Tu-144. They reported that the aircraft did not seem ready for scheduled flying and apparently had been pushed into service with great haste, a strange hiccup as this showed even in small items, that presumably could have been remedied fairly easily: several panels in the cabin ceiling were half open, some fold-down tables were jammed, window blinds closed randomly by themselves while some reading lights and lavatories also didn't function.

After exactly two hours, the inaugural flight landed in Alma-Ata. Then a long wait began, as the aircraft was pushed back and forth in the attempt to position it at the exact position of the gangway. Only then could the big reception begin with workers waving flags and displaying banners, a carefully staged event, the cheering crowd lined up with the impressive mountain scenery of Alma-Ata airport towering from behind. Afterwards, the inaugural guests flew back to Domodedovo on the same aircraft, Tu-144 registered CCCP-77109, where they landed safely six and a half hours after departure. Captain Boris Kusnezov was reasonably ecstatic for the occasion: "Everything worked perfectly." The passengers, however, had rather mixed reviews: "The feeling of acceleration at takeoff is extraordinary," found Reuters correspondent Charles Bremner.

"It felt as though we were being ticked in the back, and the ground just fell away beneath us as we took off." Daniel Vernet of *Le Monde* described the flight as "perfectly smooth", while he was pointing out a shortcoming the Tu-144 never overcame: "During the flight the cabin is noisy. One can have a conversation only with difficulty." Howard Moon described it as "a cacophony only suitable for combat crews (equipped) with helicopter sound-proofing helmets."

Aleksey Tupolev, who was on board the first flight, promised: "We are working on the noise problem." But he never delivered solutions due to scarce funding, the noise of the engines, the heat protection system (working with used cabin air) as well as the air conditioning remained infernal. Passengers seated together could only communicate by screaming, while if they sat even one seat apart, writing notes remained the only option. The worst affected zone was the aft part near the engines, as the average noise level in the entire Tu-144 cabin was already up to 95dB, whereas it is about 85dB in today's aircraft cabins. To lend sound-absorbing helmets to passengers never crossed Aeroflot's mind, though it was often suggested. Aeroflot pointed out the route had been chosen for supersonic flying, as it only crossed scarcely populated areas. No one ever further elaborated on the topic

of the sonic boom and flights over land were always planned in the Soviet Union. After the premiere, Aeroflot put on a supersonic flight every Tuesday. Flight SU499 was due to depart Moscow Domodedovo at 8.20am, arriving at Alma-Ata at 10.50am, while the return flight was designated SU500; both were two hours faster than the subsonic flights between the two metropolises. Although at the time eight serial aircraft had been assembled, with two spare ones parked nearby very visibly during the ceremonies at Domodedovo, the plan remained to just put on one round trip per week. The start was a bit rough, though, as the first two Tuesdays, the Tu-144 flights were cancelled, whereas on the third they were postponed. By December 27th, 1977, the first serious of many technical problems to come during flights occurred, with a cabin depressurization.

That is why the supersonic airliners in their rare deployments soon earned the name "fair weather jets", as each flight was a nail-biting affair for the ones responsible.

"Among the crews there was a saying: 'Flying the Tu-144 is like kissing a tiger'," recalled Vladimir Potemkin, deputy head of operations at the Ministry of Civil Aviation. "There were no fuel reserves in case the destination airport Alma-Ata would have been closed, and

The view of the Tu-144's underside reveals how in the production version, the engines were integrated between fuselage and wings to enhance aerodynamics (AS)

if then Tashkent as the only alternate was unavailable, you couldn't land anywhere. For that reason, for every flight the entire Ministry was on alert, every fifteen minutes we inquired about the weather at both airports." A procedure was established that not one, but two deputy Soviet ministers, the one for civil aviation and the one for aerospace industry, had to give a green light for each single scheduled flight. As had Aleksey Tupolev, this all resulting in many cancellations at short notice. This demonstrates how

high the upper echelons in the Kremlin saw the risk for a catastrophe during supersonic operations. And this worst case already seemed inevitable on January 25th, 1978, when 22 of 24 essential systems on board failed. As many important media and guests were on board, it was decided to go ahead with the flight, despite eight systems having failed already before takeoff. In the air, the situation had become so dramatic that even Kremlin leader Leonid Brezhnev was informed due to the expected bad publicity of

the looming catastrophe. In the end, miraculously, all went fine. Despite the many incidents, Howard Moon assessed: "It is remarkable that this incarnation of 1960s technology and values got this far. The Central Committee's launch of Tu-144s 'scheduled service' represented the triumph of hope over experience. Launching this propaganda vehicle for dubious and limited dividends, given Soviet technical inadequacies, reveals the quality of Soviet decision making in general."

While "scheduled operations"

initially ran at the lowest possible level and with all sorts of safety nets in place, the Tu-144 program soon took another blow in May 1978 when the first production aircraft of the new long-haul version Tu-144D was on a test flight prior to delivery to Aeroflot. It was equipped with the more powerful RD36-51 engines by Koliesov, on which they had worked since 1964, which only finally flew a decade later in November 1974. The engine delivered 20 tons of takeoff thrust and 5.1 tons of thrust in supersonic cruise, more than the intended 1.23 kg. A further increase to up to 24 tons of takeoff thrust was planned. Only by this time was propulsion of the Tu-144D finally getting close to the performance of Concorde's Olympus engines: in June 1976 a Tu-144 carrying five tons of payload covered a distance of 6200 km.

But even before delivery to Aeroflot, a serious accident occurred on May 23rd, 1978. Under the command of Eduard Yelyan, who had also operated the first flight, strong vibrations caused a leak in a fuel line, leading to eight tons of kerosene accumulating in the nacelle of engine number three. The flight engineers deemed the dwindling fuel reserves indicated by their instruments to be faulty and did not report them to the pilot. But soon a fire broke out and the pilots cut off engines three and four. Then

on top of that, engine number two failed and the aircraft started trailing fire, the cockpit filled with smoke. Still the crew succeeded in a belly landing on a field close to Yegoryevsk, over a hundred kilometres southeast of Moscow. In it, the aircraft buried its torn-off droop nose underneath the fuselage, the nose puncturing the cockpit at exactly the point where the flight engineers were sitting who subsequently died. The technical trigger of the accident had to do specifically with the Tu-144D's fuel pump system and no connection with the Tu-144S flying passenger services. But Minister Boris Bugayev had had enough of it now, with the actual number of serious incidents probably being even higher than the known ones. He decreed the immediate termination of Soviet supersonic passenger flights. The 55th and last one took place only a week later, on June 1st, 1978. On all passenger flights, just 3194 passengers had been carried, an average of 58 per flight. But there were still plans to restart commercial services with the newest version of the Tu-144 between 1980 and 1982.

From June 23rd, 1979, the improved Tu-144 was deployed on a cargo-only Aeroflot service from Moscow to Khabarovsk with speeds of up to Mach 2.15. But again, another serious incident happened on August 31st, 1980. The Tu-144D registered CCCP-77113

The second test aircraft of the Tu-144 in Moscow before going to the Paris Air Show in 1973 (AA)

suffered an uncontained engine failure flying above Mach 2 at about 16,000 metres of altitude, a compressor stage burst and damaged the fuselage and wings. The aircraft went into a steep dive before pilot Yevgeny Goryunov managed to regain control and resort to an emergency landing at Engels 2 military base close to Saratov in southeast Russia. After that, there were few activities with the Tu-144, as drastic increases in energy prices had also reached the Soviet Union, letting fuel guzzlers such as civil supersonic airliners become fairly obsolete in a common effort to save costs. Including

the "scheduled" mail and cargo flights between December 1975 and April 1978, there had only been 103 commercial flights in total, for which the aircraft had been flying 181 hours, 104 of which at supersonic speeds. During these very limited operations, however, a total of 226 incidents or failures had occurred on board, 80 of them in flight – at least this was the number the Soviets reported to the UK when they asked for support.

On January 7th, 1982, the end of production for the Tu-144 was announced. Up until then, a total of 17 aircraft had been built, the prototype, two pre-

production- as well as 14 serial aircraft, of which nine were Tu-144Ss and five Tu-144Ds, one of them remaining unfinished. On July 1st, 1983, there was an official government decree to cease the program and use the remaining aircraft as flying laboratories. In the same month, Tu-144 registered CCCP-77114 was still doing test flights, breaking 13 world records. One of them a circuit over 2000 km, was flown at 18,200 metres carrying 30 tons of payload with a minimum speed of 2032 km/h. Many of these records, however, have later been surpassed by Concorde. In 1985, the last Tu-144D built was used in training

In the sudden political thaw between East and West, American and Russian partners agreed on a joint supersonic test program in 1994 using the Tu-144LL (AA)

Veterans in hibernation: Tu-144s CCCP-77115 and RA-77114 stored in Moscow Zhukovsky in 2018. The former NASA test aircraft was placed prominently on display near the terminal in 2019 (AA)

A total of 26 test flights doing basic research for future civil supersonic projects were operated jointly by NASA and its Russian partners from 1996 to 1998 on the Tu-144LL (AA)

crews for the Russian space shuttle Buran – which can be seen at Technikmuseum Speyer in Germany, whose other branch in nearby Sinsheim is the only place worldwide displaying a Tu-144D alongside Concorde. Between 1986 and 1988, Tu-144D registered CCCP-71114 was also used for medical and biological tests of cosmic radiation and other scientific tasks before it was mothballed in 1990.

After the fall of the iron curtain and the end of the Soviet Union, something unexpected happened in the early 1990s: US Vice President Al Gore and Russian premier Viktor Chernomyrdin signed a cooperation treaty in 1994 agreeing on joint research in the field of supersonic airliners. That enabled the re-activation of the Tu-144D now registered RA-77114. At the time she only had 82 flight hours behind her. In an unprecedented Russian-American cooperation NASA, Tupolev, Rockwell and later Boeing gathered to modify the Tu-144 as a test vehicle for a research program aimed at a future SST. Investments by the Americans of US$19 million were necessary, plus an unknown sum spent by the Russian side. Due to the major modifications, the aircraft received the designation Tu-144LL, LL standing for the Russian acronym for flying lab. The aircraft was equipped with new Kusnezov NK-321 engines, having been developed for the Tu-160 *Blackjack* bomber. They offered 25 tons of takeoff thrust with afterburners ignited and increased speed to Mach 2.3 at a range of 6500 km – parameters the Tu-144 could have only dreamed about in her short operational life as airliner. Only this engine was now, after the Tu-144's career had ended,

Yes!

Specifications

Number of crew   3
Overall span   90.7 ft.
Overall length   191 ft.
Cruising speed   1560 mph.
Range   4560 mi.

Payload   26.400 lb.
Runway length required
according to ICAO
regulations=8850 ft.
Passengers   120

AVIAEXPORT

An ad from the early 1970s when the Soviet hoped to export the Tu-144 (AS)

capable of cruising at Mach 2 without ignited afterburners. At the same time, the flying lab was equipped with digital devices for data recording and a wide array of measurement equipment and microphones, plus emergency bailout options for the crew. Between November 1996 and April 1999, a total of 27 test flights were performed, 14 of them supersonic, before the program was cancelled due to lack of funding. The final flight of a Tu-144 ever took place as part of the research program on April 14th, 1999. Since August 2019, the former flying lab is prominently displayed in front of Zhukhovsky airport.

In the end, the summary of the Tu-144, the world's first supersonic airliner, is sobering. The total cost of the Concorde program is estimated at US$25 billion today, and while there is no information on the cost of the Tu-144, it was likely lower. A bad investment in any case. Howard Moon quoted unidentified Western experts in 1980s Moscow as saying: "In terms of investment and return the Tu-144 may well rank as the biggest single failure in the whole history of aviation." It is almost tragic that the immense and finally unsuccessful efforts to gain a share of supersonic air traffic, led Soviet and today's Russian civil aircraft manufacturing to become what it is: completely irrelevant. Alexander Pukhov, later one of Tupolev's leading

designers, said in 1998: "In my opinion, this was an aircraft then that was ahead of its time and the country's capabilities by ten to 15 years." Howard Moon dedicated much space in his book to the Soviet failure and sums things up in a very pointed way: "Seven years of design and study work were followed by fifteen years of flight testing. Two major designs were developed that involved at least four major wing designs and two different engines, fixed in four locations. Yet for all their efforts, in the end the Soviets were left with a rugged, powerful airframe, noisy and gluttonous of fuel, unrefined beyond the level of a high-performance military interceptor." And he goes on analysing the motives behind the Soviet SST: "The Tu-144 saga shows the powerful appeal of the myths of power, speed and futurism. … The Tu-144 was always an anachronism, born out of due time, more a realized wish-fulfilment fantasy than a reflection of reality. … Noisy, inefficient, elitist – it was an embodiment of yesterday's tomorrows."

Tu-144 CCCP-77106 last flown in 1980 is now on display the Russian Central Air Force Museum in Monino near Moscow (Andrey Khachtryan)

five

Is there a market for commercial supersonic flight? Myth and reality

"To spend many billions of dollars of the taxpayer's money on a project that is claimed to be a commercial possibility is inexcusable. Because it deliberately ignores the elementary requirement for any commercial project, be it cornflakes or SSTs, that there should be a market for the product."

Ronald Davies in "Supersonic (Airliner) Non-Sense", 1998

British Airways Concorde approaching Barbados (AS)

Until this day, the assumption that one could make a profit by operating passenger flights with at least double the usual cruise speed has consistently turned out to be an illusion. Any project in this respect, either in the planning stage or operational, as the previous chapters have shown, has been a financial disaster. The respective gargantuan efforts of the world's biggest economic powerhouses to establish supersonic passenger operations and build the aircraft needed for them have led to nothing in the end. A market that was worth being called a market never existed. Even if all nations involved so far disputed this claim vehemently. Only for them to end up paying the astronomical bill in the end, without getting any equivalent value in return except for elegant aircraft serving as eye-catchers in museums and some technological experiences being used as spinoffs elsewhere later. But without finding any commercial demand to rectify such operations there has been no hint of possible profits to be made. In other words: the idea of commercial supersonic operations is an illusion. That has always been the case and chances are it will stay like that. Even though the makers of the next generation of attempts to establish supersonic passenger services, being portrayed in this book, now again claim the opposite.

But can they turn history around as well as ignore undeniable facts?

A thorough and critical look is in order into the realities of this perceived market niche, of which some will remain simple facts, even with the newest technologies and possibilities of the 21st century. There was a wise book written about this topic, published in 1998 by Ronald Davies (1921-2011), who spent almost all of his life in the aviation industry, working for aircraft manufacturers, ministries, airlines and for 30 years as a curator at the Smithsonian Air and Space Museum in Washington DC. His work "Supersonic (Airliner) Non-Sense" remains a brilliant pamphlet even today, which anyone should be forced to read who is contemplating getting engaged in the supersonic "business". And first of all, everyone should embrace Davies' mantra: "No more than half a dozen city pairs could ever justify supersonic flights."

It is almost amusing to rediscover the near-religious fervour with which, for example, Henri Ziegler was defending the Concorde programme and its supposed market in June 1971 at the Paris Air Show. Henri Ziegler at the time was President of Sud Aviation and head of the Concorde programme, later one of the co-founders of Airbus Industrie. As it looked possible that Concorde might

not be welcome in the US due its noise levels, he explained in an interview with German newsmagazine *Der Spiegel*: "The North American market is not the only one for air traffic. There is a very important market to South America, Africa, the Far East, Japan, China, Australia and India." Apparently hinting at markets for supersonic flying, because that was the only topic of the interview. As the journalist objected that the North Atlantic was by far the most important market, without

which Concorde could never fly economically, he disputed even this assertion that is truly hard to deny. "That isn't clear at all. We have a great development before us. Fast aircraft have just changed relations between very distant countries significantly. An example: Still today, the economic relations between Europa and Australia are very insignificant because of the great distance. If this distance can be reduced by half, which is very important, the economic

relations would significantly grow."

The assertion that extremely long routes like between Europe and Australia could be meaningfully served by supersonic aircraft was lacking any practical significance even then, as Ronald Davies takes pleasure to prove. Initially he describes how the focus in marketing Concorde was placed on long routes such as from London to Australia. With ground times of 50 minutes each at two fuel stops necessary in Bahrain

Seattle—Tokyo
Subsonic—9 hrs. 41 min. non-stop   SST—4 hrs. 48 min. one stop

New York—Rio de Janeiro
Subsonic—9 hrs. 13 min. non-stop   SST—4 hrs. 45 min. one stop

New York—Paris
Subsonic—7 hrs. 3 min.   SST—2 hrs. 49 min.

New York—San Francisco
Subsonic—5 hrs. 8 min.   SST—2 hrs. 20 min.

Chicago—Los Angeles
Subsonic—3 hrs. 34 min.   SST—1 hr. 46 min.

New York—Los Angeles
Subsonic—4 hrs. 54 min.   SST—2 hrs. 14 min.

Already in 1966 Boeing tried to tout the time gained on supersonic flights. Much of this never became reality, as flying overland was banned (AS)

Four stops to Sydney: The globe-spanning supersonic network from a Concorde marketing brochure of the 1960s bears little realism from today's standpoint (AS)

and Singapore, a Concorde would have landed in Melbourne after an elapsed travel time of 13 hours and 15 minutes, a gain of ten hours versus the subsonic flights offered at the time. But this was only achievable if there were no significant restrictions in flying supersonic over land, but restrictions were foreseeable then. And by 1998 it was already possible to cover the distance from London to Sydney on a scheduled Boeing 747-400 service in 20 hours and 40 minutes with just one stop in Singapore. So even then, a supersonic flight on the route taking 13 hours was obsolete, especially with several fuel stops and a cramped cabin, which wouldn't have been appreciated by passengers. "This was still a long flight, with little chance to sleep, because of the four-hour-segment times," Davies remarks, "and the speed did not solve the jet lag problem, as was sometimes alleged".

The author of this book can attest to this: after flying on Concorde from London to New York he was attending a Broadway show in the evening local time, fighting hard against sleep – despite having trusted the British Airways advertising claim that Concorde would

COMPARISON OF FLIGHT TIMES	DISTANCE		SUBSONIC AIRLINER	CONCORDE	COMPARAISON DES TEMPS DE VOL
	STATUTE MILES	KILOMÉTRES	HRS. MINS.	HRS. MINS.	
LONDON — NEW YORK	3,580	5 750	7 35	3 17	LONDRES — NEW YORK
PARIS — NEW YORK	3,770	6 050	8 00	3 25	PARIS — NEW YORK
PARIS — DAKAR — RIO DE JANEIRO — BUENOS AIRES	7,270	11 700	14 30	7 15	PARIS — DAKAR — RIO DE JANEIRO — BUENOS AIRES
LONDON — BEIRUT — BOMBAY — MADRAS — SINGAPORE — SYDNEY	11,490	18 500	23 00	11 20	LONDRES—BEIRUT—BOMBAY—MADRAS—SINGAPOUR—SYDNEY
NEW-YORK — CARACAS — BUENOS AIRES	5,510	8 850	10 45	5 15	NEW YORK — CARACAS — BUENOS AIRES
SIDNEY — MANILLA — TOKIO	5,990	9 620	11 45	5 40	SYDNEY — MANILLE — TOKIO

minimise jet lag. But it did not, as the human body cannot be easily outsmarted. And Australia as a potential market is purely wishful thinking in any case: "With the urban population of the whole Australia no more than about six million,

and fragmented, there was insufficient traffic potential to justify the high-fare and costly operation," Davies asserted. Time and again, route categories such as Europe to Australia or Europe to South Africa were praised by market projections

as ideal for supersonic flights, according to Davies. It often appeared to him as if people in charge of such predictions did not study elementary geography at school, as they chose to ignore simple demographic realities. "The total population

of Australia is less than that of Greater New York, and the total population of all the main cities of South Africa is less than that of London." And where there are no or few humans, there are no markets. But this still didn't hinder the next but one

Is there a market for commercial supersonic flight? Myth and reality

67

generation of supposed market researchers at the beginning of the new millennium drawing routes to Australia as allegedly lucrative future markets for a possible new generation of SSTs into their colourful marketing brochures.

In the early 1970s, the potential for supersonic flights still seemed unlimited, but the naivety was as well, as to how efficiently they could be operated. "Extensive market studies come the conclusion that Concorde will have a significant part of the market with people who value time on long haul travel, and that there will be a price parity with subsonic flights," claimed Henri Ziegler in talking to *Der Spiegel*. "We estimate that Concorde flights will come at a 25 to 30% surcharge on current tourist class fares," Ziegler asserted and stressed that this wasn't overly optimistic, but "the result of our investigations". The exploding costs, which made this appear all but impossible, appeared as "embarrassing" to Ronald Davies early on. Besides the huge efforts in research and development until production maturity, Concorde's engines were mostly to blame. Because they "gulped fuel as a dehydrated survivor from drought gulps water," as Davies put it.

And added to this was the naivety of the engineers in believing they could soon overcome the sonic boom, and that of the marketing people in thinking that they could simply ignore this problem. "It was realized that an aircraft carried its boom with it along a swath of affected territory all the time it was flying supersonically, and not, as many thought, only when it actually penetrated the so-called sound barrier," wrote Davies. "Not to worry, the engineers and project managers and aerodynamicists said. This is simply another technical problem that will be overcome, just as we have solved all the other problems in the past. But it was not solved. The sonic boom is a fundamental physical phenomenon, and you can't fool with Mother Nature." Henri Ziegler's statement in this respect in his *Spiegel* interview is almost negligently naïve and totally ignorant of the facts, even in 1971: "Concerning the sonic boom, it is premature to claim that a supersonic aircraft cruising at high altitude will at all be representing a major impairment. One thing is sure for certain: There is such an important market for flights above oceans and unpopulated areas that one can avoid crossing inhabited areas." Ronald Davies, however, reported that it was already conceded in the UK in the mid-1960s that there was not any remedy against the sonic boom. "That's what reduced the market abruptly and devastatingly to just overwater routes."

But Henri Ziegler was not going to accept this by any means. A critical debate about the project, he asserted, might have made sense in 1962, but "in 1971 it doesn't make sense anymore". Concorde had become too big to fail, as the Americans say, such a gigantic undertaking couldn't be let to go bust. "The major part of the work has been already done, and to halt the program today would be a monumental stupidity, consisting of throwing enormous efforts and a very major success into the garbage and leave the monopoly to others," insisted Ziegler. To

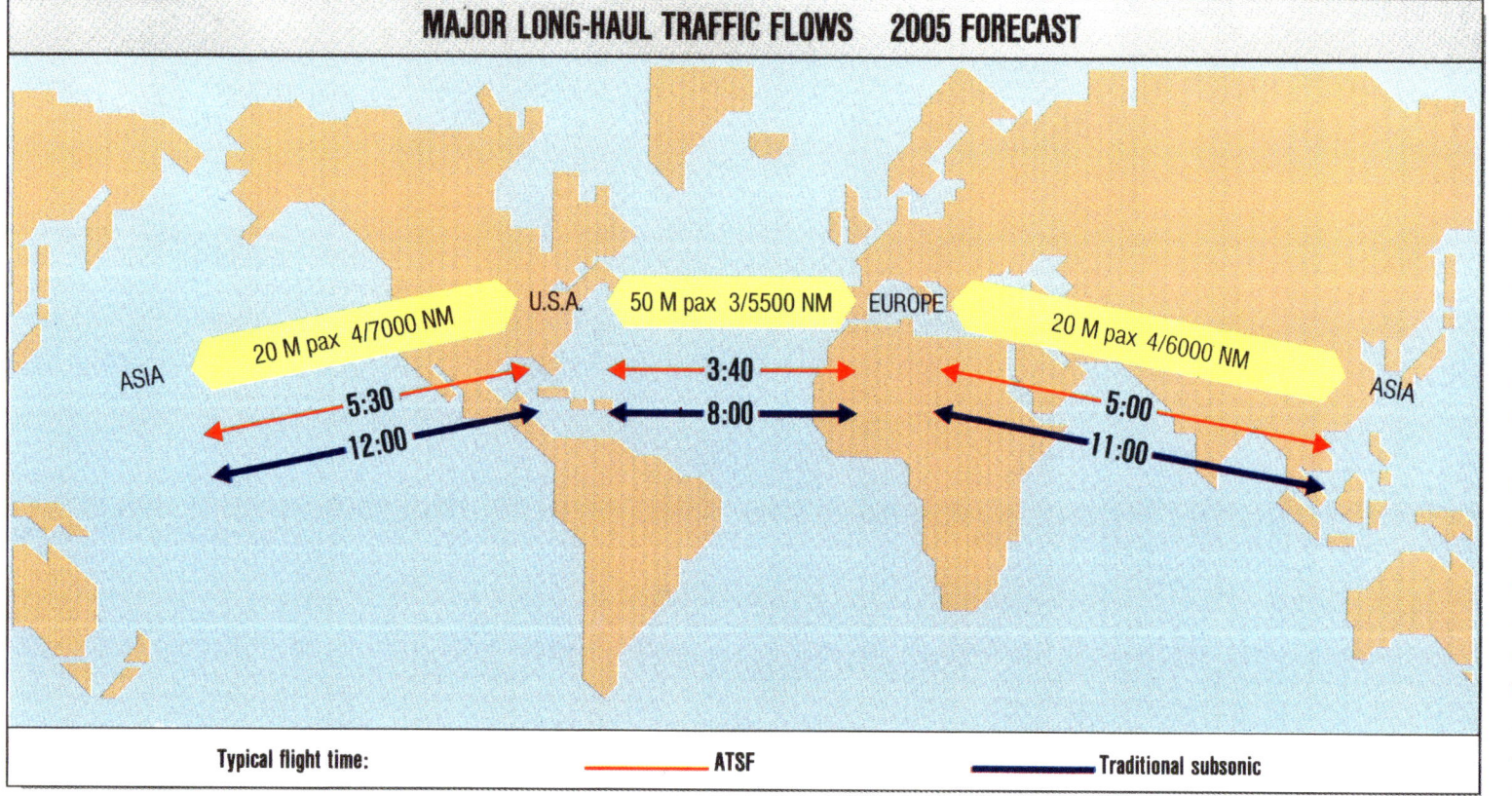

A 1990s promotional brochure by Aérospatiale not even attempted to outline the share of its ATSF SST design of the overall market expected for 2005 (AS)

their advantage, the Americans had cancelled their Boeing SST programme at almost the same time, despite huge investments. Only the Tu-144 remained as a potential competitor, but in it the Concorde programme director saw only a luxury problem, if at all. "I believe the growth of this market will be very quick and our two aircraft will complement each other. We might have difficulties to even satisfy market demand. You see how much I believe in the development of this market," stressed Ziegler.

We also see how dazed by utterly unrealistic scenarios the highest echelons of the aerospace industry and politics were at the time. Almost narrow-mindedly and stubbornly, Ziegler refused to let anyone taint his mantra, which he repeated over and over again in the *Spiegel* interview and summed up as follows: "Concorde is an excellent program fulfilling a remarkable goal," he stressed, boiling it down to the simple formula: "Concorde is a success and supersonic flight a necessity." On January 31st, 1973, at the latest, Ziegler's dream was over: Pan Am cancelled its options and down payment for seven Concordes, made a decade earlier against the great resistance of the Kennedy administration. The airline explained its decision was due to the shortcomings of Concorde, offering significantly less range and payload capacity

but higher operating costs than the current and future wide body aircraft, which would lead to higher fares than those at the time. This meant "the aircraft doesn't fulfil the future requirements the company sees today". Kenneth Owen called this the "collapse of the worldwide airline market for Concorde". Or, as Ronald Davies commented: "Without Pan Am, it was dead." And with this, any remaining illusion that there could ever be economical supersonic flight operations should have been put to rest once and for all.

One of the most substantial arguments supporting the assumption that there is not a real market can be understood by doing a very simple analysis, yet for decades generations of market researchers seem to have completely ignored it. Two of the world's major metropolises are New York and Tokyo; both are global financial centres. Which is significant, not least because bankers are a core clientele for supersonic flights, and so should be essential in any market calculation. The great circle distance of almost 11,000 km between the two cities is too far to be flown nonstop for any SST ever conceived. Meaning the necessity of at least one refuelling stop on the way and thus an unwelcome extension of travel time. Independently of that, one only has to look

at the potential flight duration between both cities and the time zones they are located in to quickly understand: there will never be a real market for supersonic business trips here. And without the clientele traveling on corporate expense, supersonic flights will never be economical. Ronald Davies was demonstrating these simple facts in easy-to-digest graphics. "These are the fundamental problems of supersonic flying over the Pacific," according to Davies, that also apply to routes eyed by the most recent supersonic projects such as Boom, for example between San Francisco or Los Angeles and Tokyo.

New York and the Japanese capital are usually apart by ten hours of time difference, "this negates the advantage gained by speed," Davies stated. "Passengers arrive at the destination either in time for a day's work only after already being awake during part of or a whole working day. Or in time to prepare for bed, after being awake for about 24 hours." Both are unrealistic and most of all, not desirable for anyone. Two examples: leaving New York during lunch time at 13.00 hours local and flying supersonically to Tokyo with twice the speed of an ordinary airliner takes only half as long, including fuel stops. On average probably about seven hours westbound. Arrival in Tokyo would be at 10am in

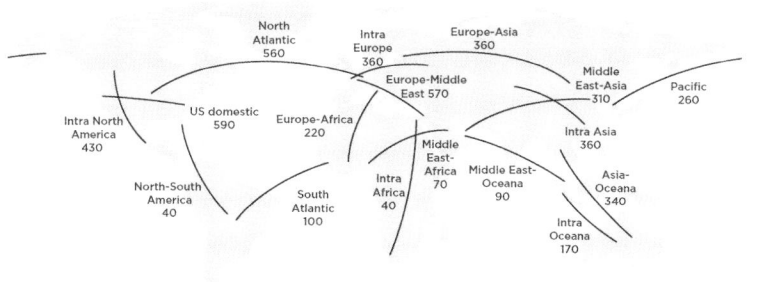

Figure 2. Daily commercial supersonic flights by market

The international climate research group ICCT even in recent times used totally unrealistic assumptions of how the number of daily SST flights could look in 2035 (ICCT)

the morning local time. A good night's sleep like in a lie-flat seat in First or Business Class of a subsonic aircraft would be unthinkable in a cramped cabin of an SST. Even more so as for the body clock, the daytime hours of the flight would not be sleeping time, and in addition the fuel stops would interrupt resting. Leaving New York in the evening local time does not fare much better, as arrival in Tokyo would be in the afternoon or early evening, when business meetings do not make much sense anymore. In the opposite direction, from Tokyo to New York, flight time is even shorter thanks to the jetstream, Davies calculates about six and a half hours flight duration by SST. But the problem remains: Leaving Tokyo at the local lunch hour means arrival in New York in the early morning. Departing Japan in the evening leads to arrival in Manhattan in the early

afternoon.

"It is bad enough having to face razor-sharp Japanese businessmen even after a good night's sleep, early morning workout, and bacon and eggs for breakfast," wrote Davies. "But to do so fully jet-lagged (for the human constitutional clock does not adjust itself because the body goes supersonic) at the equivalent of one o'clock in the morning is not the best way to start a negotiating day. If the answer is to take a rest, either before or after the flight, then the whole objective of supersonic travel is negated," judged Davies. And he went to great lengths to prove that this wrong assumption flawed all calculations about Concorde's expected customer base. Even on the link between Europe and New York, which is more suitable for supersonic travel in all parameters than transpacific routes. By 1990, one-way tickets

Is there a market for commercial supersonic flight? Myth and reality

69

PROBLEMS OF RANGE, SUPERSONIC FLIGHT OVER LAND, JOURNEY TIME, AND JET LAG

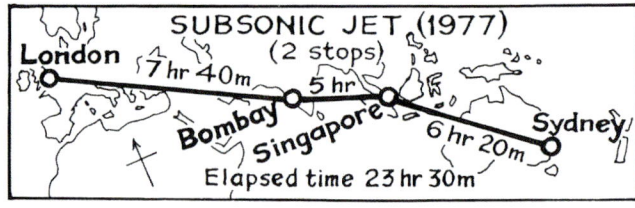

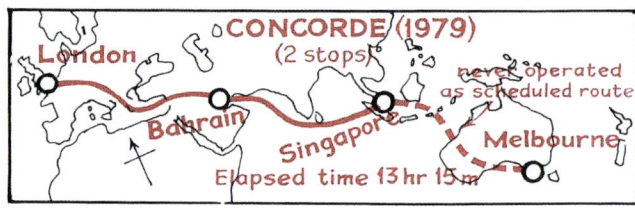

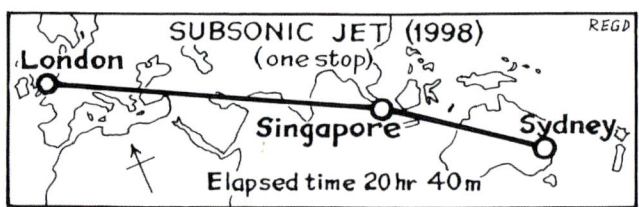

One of Ronald Davies' graphics explaining how little sense a supersonic connection from Europe to Australia would have made (Paladwr Press)

on Concorde from New York to London already cost about US$4000 (roughly double as much in today's value). And such fares were paid by very few US companies even then, not even for high-ranking executives. Why pay so much more in order to arrive a few hours earlier in London or Paris at a local time when the working days had ended there anyway? Another wrong assumption was that the entire super premium customer base would fly Concorde. Even when traveling between Boston or Philadelphia and Frankfurt or Zurich the direct service in a subsonic airliner was not only more comfortable, but also quicker and more reliable in the end, as one couldn't miss any transfer connection between hub and final destination.

The author of this book has confronted Blake Scholl, CEO of Boom Supersonic and closely involved in building the next generation of supersonic airliners, with the critical points that Ronald Davies published over two decades earlier. And, probably predictably, Scholl answered: "I don't understand the argument that supersonic flight times don't work across the Pacific. There are plenty of examples that do, including New York to Tokyo." Blake Scholl, however, bases his assumptions on a speed of Mach 2.2 that the *Overture* SST Boom Supersonic wants to build is supposed to reach, faster than Concorde. Scholl gives the following examples:

* Depart San Francisco at 8am, arrive Tokyo 6:15 hours later at 7.15am Tokyo time. Spend a full day in Tokyo, catch an evening overnight flight back. You just had a full working day in Tokyo with 24 hours away from home and arrived back in the States before jet lag set in.

* Depart Sydney at 8pm, arrive Los Angeles eight hours later at 9am, a quite reasonable "redeye" night flight. There are even plenty of great supersonic flight times between Tokyo and New York. At Mach 2.2, you can depart Tokyo at 6 pm and arrive in New York 9:5 hours later at 1.30pm - a fairly reasonable "redeye". From New York, you can depart at 11am and arrive Tokyo at 10.30am. You can argue that's a long day, but it's still meaningfully better than today's subsonic flights that don't arrive until 3pm.

You can argue New York to Tokyo supersonic is similar to London-New York today, which of course is a great market.

Scientists cannot share the newly emerging euphoria as exuded by Boom. "Part of the full truth are the transfer times to, from and within the airports, in the best case four times one and a half hours each," calculates Bernd Liebhardt, supersonic expert at the German Aerospace Centre (DLR) in Hamburg. And he comes up with further reasons for scepticism: "The fuel stops outbound and inbound between North America and Asia over the Pacific. And the order to put 'seats upright' cause people to wake up and prevent a reasonable night's sleep." In addition, "seats are the size of Premium Economy and not really suitable for relaxed sleeping" and "the unusual work times at destination". Therefore, explains Liebhardt, "I don't even look at flights with intermediate stops in my market studies".

It appears that again, the same well-known mechanisms are at work here, with the advocates of a new supersonic era sweet-talking and creating an imaginary market, just the same procedure as the Concorde makers used in the 1960s and 1970s. Of course there are currently still many unknowns in all these calculations for a new supersonic era. For instance, the question of whether it will be possible to offer supersonic flights at the same fare levels as today's Business Class tickets on identical routes. And if *Overture* can indeed fly faster than Concorde. Ronald Davies

already tried in 1998 to paint a future scenario as realistically as possible for one of the NextGen SSTs discussed then. The assumption was for a 300-seat SST and he predominantly looked at transatlantic as the world's most important long-haul passenger market in the world. For 2010, he saw as the four biggest SST routes as being London to New York with 850,000 SST passengers a year, followed by London-Los Angeles (270,000) and London-San Francisco (160,000) before Frankfurt-New York (144,000) as the only route from the European continent among the top five. A fifth of that route would have to be flown over land and therefore below Mach 1 and the route would only utilize one supersonic airliner per year. For the entire top ten transatlantic routes Davies saw a need of just 16 to 17 SSTs. "Add a half dozen more for a possible Pacific market and throw in a few more SSTs for luck, with all the factors of special pleading and wishful thinking, and the world market for an SST can never be more than two or three dozen." The question remains whether anyone is interested in these very sober, but most likely accurate analyses. Or whether the well-known reflex can be detected again of employing wishful thinking to make up a market which lets their own product appear desirable and useful.

Braniff flew Concorde on domestic US routes briefly, in extension of transatlantic flights by BA and AF, but it was never painted in their livery nor was there a sustainable demand (Artists impression)

It is not known if the analysts of Swiss bank UBS have ever cared to read Ronald Davies's insights. Looking at one of the first new, thorough market analyses for SSTs in a long time, published by the bank in late 2020, it does not seem likely. UBS sees a supersonic market emerge only around 2040 with a potential US$160 billion supersonic aircraft market, worth US$85 billion for SST business jets such as the proposed and now terminated Aerion AS2 project, and US$75 billion for commercial transports like *Overture* by Boom. For commercial operators (airlines and charter companies) there will be a cumulative market potential of US$180 billion, adding up to a total supersonic air travel market volume of US$340 billion, according to UBS. Their analysts also conducted a poll among potential travellers, of which 20% said they would pay at least 50% more for a ticket if travel time could be cut in half. Which is precisely not the business idea of Boom, who wants to offer supersonic travel at par with today's Business Class fare levels. At least the bank's analysts ask a valid question: "Can supersonic travel mesh with an increasingly environmentally-focused marketplace?" To then assert: "The new supersonic jet players are targeting messages around sustainability merits of their offerings, but the greening of aviation will be a counterweight to faster adoption." Which, as the authors rightly point out, is not a new thing after all. While aviation overall did not run into major confrontation with environmentalists until the 1980s, the proposed US SST saw fierce opposition due to its impact on the environment in America in the late 1960s, as did Concorde in New York prior to its arrival there in the mid-1970s.

According to the analysis, the bankers see 1.6 million flights carrying 375 million passengers on 2,300 route pairs that could potentially be replaced by SSTs of the kind currently planned. As an initial demand for commercial supersonic airliners the analysts see 170 aircraft, growing to 400 by 2040 at prices of US$200 million per aircraft. Readers who have embraced Ronald Davies's emblematic warnings should see a red light of caution flickering here, as it is a weak foundation the analysis is based on. In general, supersonic business jets are seen as the most likely variant of any future highspeed passenger flying, and this is reflected by UBS's assumptions. At first the authors see a demand of 45 SST business jets a year, then 90, accumulating into a total fleet reaching 700 aircraft by 2040, costing US$120 million per piece on average. Whether Ronald Davies' predictions will retain the upper hand posthumously remains to be seen. In any case, commercial supersonic flying always was and always will be a difficult market, to say the least.

Is there a market for commercial supersonic flight? Myth and reality

71

six

Miracle with delta wings – history and design of Concorde

"Concorde has given the European aircraft manufacturing industry a technical achievement … almost comparable with those earned in the United States and the Soviet Union by their space efforts."

Ronald Davies, 1998

Concorde during engine tests in Toulouse, ca. 1974 (Airbus Heritage)

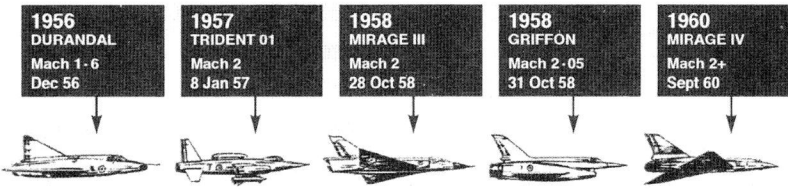

1956 DURANDAL Mach 1-6 Dec 56	1957 TRIDENT 01 Mach 2 8 Jan 57	1958 MIRAGE III Mach 2 28 Oct 58	1958 GRIFFON Mach 2·05 31 Oct 58	1960 MIRAGE IV Mach 2+ Sept 60

EVOLUTION OF THE S.S.T.

French supersonic experience is already extensive and includes military aircraft which, since 1957, have been flying at about twice the speed of sound.

This supersonic experience began with the Durandal and the Trident of Sud-Aviation and was followed by the Griffon of Nord-Aviation and the Dassault Mirage 3 and 4.

Since the end of 1956 French official facilities have been deployed on an examination of a supersonic airliner. Just afterwards, Sud-Aviation also began similar studies, and in 1959, the French Air Ministry officially asked Sud-Aviation, Nord-Aviation and Generale Aeronautique Marcel Dassault to work on a project for a medium range Mach 2.2 supersonic transport.

Of the three projects which resulted, that of Sud-Aviation was chosen in October 1961 and work on the Super Caravelle followed in collaboration with Dassault. A model of the Super Caravelle was shown at the Paris Salon in 1961.

Towards the end of 1961 there began discussions about a joint programme with British Aircraft Corporation who were then working on a long range Supersonic airliner of similar configuration.

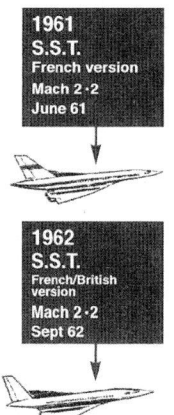

1961 S.S.T. French version Mach 2·2 June 61

1962 S.S.T. French/British version Mach 2·2 Sept 62

In 1956 the British Government set up a Supersonic Aircraft Committee which was a combined operation between the National Establishments and the aircraft industry. This Committee initiated a series of important aerodynamic and structural investigations into the shape and scope of a supersonic transport. In 1959 the Committee recommended that the British aircraft industry should make preliminary studies of an aircraft cruising at M 2.2 (some 1500 m.p.h). In 1959/60, British Aircraft Corporation was awarded a detailed design study contract of a transatlantic range aircraft to be based on work already done by Bristol Aircraft Limited, a subsidiary company of B.A.C. In France, Sud-Aviation was also doing design studies and in 1961 Sud revealed results of their work on a project bearing a close family resemblance to some of the British shapes but aimed at a shorter-range operation. At the end of 1961 Sud-Aviation and British Aircraft Corporation, with the backing of their respective governments, began to co-operate on the supersonic transport design which has now been adopted for manufacture.

EVOLUTION OF A SUPERSONIC AIRLINER

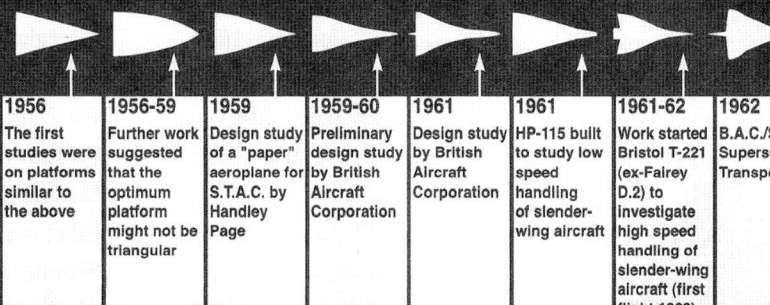

1962 T-188 supersonic research aircraft first flight

1962 Olympus Vulcan flying engine test bed

1956	1956-59	1959	1959-60	1961	1961	1961-62	1962
The first studies were on platforms similar to the above	Further work suggested that the optimum platform might not be triangular	Design study of a "paper" aeroplane for S.T.A.C. by Handley Page	Preliminary design study by British Aircraft Corporation	Design study by British Aircraft Corporation	HP-115 built to study low speed handling of slender-wing aircraft	Work started to study Bristol T-221 (ex-Fairey D.2) to investigate high speed handling of slender-wing aircraft (first flight 1963)	B.A.C./Sud Supersonic Transport

∗ In Great Britain alone some 300 models of supersonic wings have been tested in Ministry of Aviation wind tunnels. The development of the slender wing has been expedited by a co-operative effort from Government Research Establishments, the Aircraft Industry and the Universities

On February 25th, 1954, the first meeting of well-reputed scientists took place at the Royal Aircraft Establishment in Farnborough to seriously discuss a supersonic airliner. Sir George Edwards, head of Vickers Aircraft, was certain at the time that there was "definitely a link between industry, population and the speed of modes of transport". In 1956 Edwards urged the British industry to abandon all planned subsonic jets for the time being, and fully commit to building a supersonic airliner. The real birth date of what would later become Concorde came on November 5th, 1956 when the newly established *Supersonic Transport Aircraft Committee* (STAC) met in Farnborough. Members were all organisations and institutions of British aviation: representatives of all principal aircraft and engine manufacturers, aviation regulators and both the British Overseas Airways Corporation (BOAC), then handling long-haul flying, and British European Airways (BEA), dealing with short and medium haul, (these merged into British Airways in 1974). Initially, two basic designs were envisioned, an aircraft for 150 passengers achieving Mach 2 on long haul

The British view of the evolution of supersonic flight (BAE Systems Heritage)

with a sleek delta wing, also a medium haul version boasting a wing in an unusual M-form for speeds of up to Mach 1.2 and a hundred passengers.

These designs were originally conceived by Dietrich Küchemann, a German designer at Messerschmitt. He had teamed up with genius mathematician Johanna Weber at an aerodynamic test facility in Göttingen. Both met again after the war, working for the British at Farnborough, and supplied a decisive share of the concept that later would become Concorde. It was mostly Weber's achievement to correctly calculate the lift of delta wings. That "radically altered the entire understanding of supersonic flight, and many of those involved in the early stages of supersonic research considered her ideas to be both foundational and pivotal," asserted Samme Chittum in her book published in 2018. "The resulting wing, known as an ogival thin wing or ogive, arch - was not only a technological breakthrough but also an aesthetic triumph that produced one of the most beautiful aircraft ever to fly."

In March 1959, STAC finally finished its report after two years of work, which had initially been kept secret. Its conclusions were positive and encouraged the government to commission both projects. "A stage has been reached from

which the industry should start serious and detailed design studies for two supersonic airliners," the STAC report said. "1. A long range supersonic transport capable of taking 150 passengers on non-stop London to New York operations ... at a cruising speed of approximately Mach 2. ... 2. A shorter-range (up to 1,500 statute miles/2.414 km) SST to operate on Empire and European routes with a cruise speed of Mach 1.2 and a capacity for 100 passengers," the STAC report recommended. At the time, the design of a transatlantic airliner was of lesser priority, as it posed many complex problems. Overall the STAC members predicted a market for 120 to 175 aircraft, the transport ministry was even hopeful to be able to sell up to 210 aircraft. The cost, however, would be hard to calculate, warned the STAC report. In 1959 the assumption was for £127-175 million or US$375 million at the exchange rate of the time for both models, about a tenth of the actual cost in the end. Actually developing Concorde alone, without production costs, devoured about US$4 billion, equivalent to US$30 billion today, a lot more than what was spent on developing the Airbus A380 at the beginning of the new millennium.

"The committee's recommendations arrived at a time when the political climate favoured an aggressive new

Concorde
PRE-PRODUCTION AIRCRAFT
APPAREIL DE PRÉ-SÉRIE

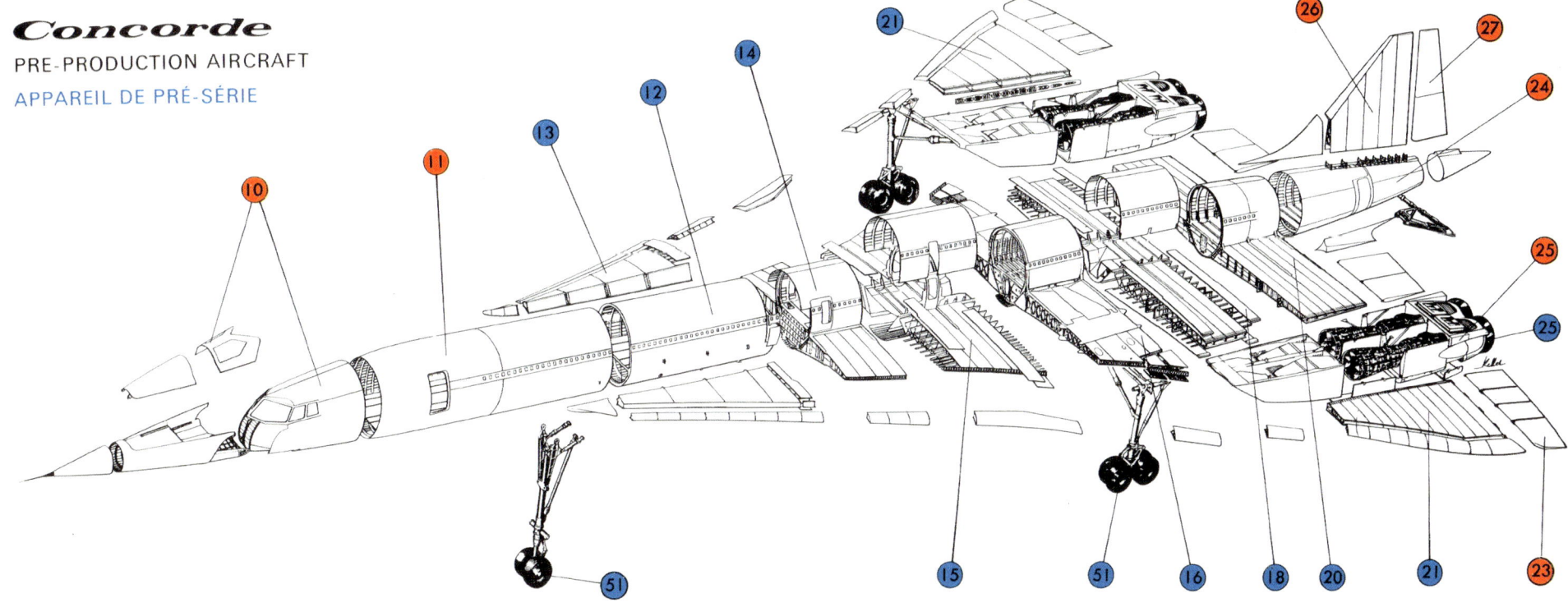

SYSTEMS   SYSTÈMES

BAC FILTON DIVISION	
ELECTRICS	ÉLECTRICITÉ
OXYGEN	OXYGÈNE
FUEL	CARBURANT
ENGINE INSTRUMENTATION	INSTRUMENT RÉACTEURS
ENGINE CONTROLS	COMMANDES RÉACTEURS
FIRE	FEU
AIR CONDITIONING DISTRIBUTION	DISTRIBUTION AIR CONDITIONNÉ
DE-ICING	DÉGIVRAGE

SUD TOULOUSE	
HYDRAULICS	HYDRAULIQUE
FLYING CONTROLS	COMMANDES DE VOL
NAVIGATION	NAVIGATION
RADIO	RADIO
AIR CONDITIONING SUPPLY	ALIMENTATION CONDITIONNEMENT D'AIR

MANUFACTURE BREAKDOWN
DÉTAIL FABRICATION

10	●	Fuselage nose	Pointe avant
11	●	Forward fuselage	Fuselage avant
12	●	Intermediate fuselage	Fuselage intermédiaire
13	●	Forward wing	Onglets AV voilure
14	●	Centre wing	Partie centrale 41 à 46
15	●	" "	" " 46 à 54
16	●	" "	" " 54 à 60
18	●	" "	" " 60 à 66
20	●	" "	" " 66 à 72
21	●	Outer wing	Voilure extrême
23	●	Elevons	Elevons
24	●	Rear fuselage	Fuselage arrière
25	●	Nacelles	Nacelles
25	●	Nozzle	Tuyères
26	●	Fin	Dérive
27	●	Rudder	Gouvernail
51	●	Landing gear main	Train principal
51	●	Landing gear nose	Train avant

The distribution of work shares in the 1960s between Toulouse and Filton gave a preview of multinational production at Airbus a decade later (AS)

Olympus 593

— POWER FOR CONCORDE

Page 42

An 1967 ad by Bristol Siddely Engines (later Rolls-Royce) and Snecma for Concorde's Olympus 593 engine (AS)

project, since British aircraft manufacturers had failed to achieve the commercial success of their American counterparts," asserted Samme Chittum. Concorde pilot veteran Christopher Orlebar analysed:

"By 1962 Britain had convinced herself that it was both highly desirable and technically possible to build an Atlantic range SST." The British attitude seemed in any case to be: close your eyes and go for it, with the principal goal of helping Britain regain industrial leadership in aerospace. "If ever there was a case of 'Don't confuse me with the facts, my mind is made up' it was the approach made by the Concorde protagonists in the 1950s," judged Ronald Davies. He called Concorde "an economic disaster from the earliest years of its embryo stage of development" and found that Great Britain has had to "pay a terrible price" for it and in its favouritism for Concorde "criminally neglected" other promising aircraft projects. "If £1 had been allocated to the Trident for every £100 thrown at Concorde, de Havilland could have snatched about a third of the Boeing 727's market," wrote Davies.

In France, serious contemplation about a supersonic airliner only began in early 1957, when the government issued a tender in the name of Air France for a jet design of medium range, capable of flying 60-70 passengers over a distance of 3500 km. Air France was keen to follow up on Caravelle's success, but was under the unrealistic assumption of being able to operate an SST at comparable cost. All three manufacturers Nord Aviation, Sud Aviation and Marcel Dassault took on the challenge and started right away to draw up concepts. In early 1960, Sud Aviation and Dassault joined together in their efforts, while Nord Aviation exited the competition. Already by mid-1961 it was becoming obvious that the explorations on both sides of the English Channel were coming to almost identical conclusions about supersonic airliners. It was also impossible to deny that one country alone would not be capable of tackling a project of such magnitude by itself. Germany and the US appeared as possible partners, but the German industry did not see fit yet or was not interested in tackling the challenge, while in the US manufacturers had already started to work on their own designs.

When a model of the French SST design called Super Caravelle appeared at the Paris Air Show in May 1961, it became obvious how close this concept was to the British study Bristol 223. But there was still a fundamental difference between both sides that caused heated arguments: the British declared that just developing a transatlantic version would be economical, while the French insisted on also building a medium haul version. In January 1962, British and French agreed to continue to pursue both ideas. The concepts on the table were identical in most parts, both in size and shape. Both had four Olympus 593 engines each, with the only difference being different fuel tanks according to the version. An internal British memo said: "There will surely be a supersonic passenger airliner, though there are doubts about its commercial and economical wisdom. Therefore we must decide if we build one, or completely withdraw from future advanced aircraft technology. If we decide in favour of it, we must not delay the decision." In the meantime, both governments were negotiating about a joint programme and 50:50 work sharing; on November 29th, 1962 this was finally announced. In London, British aviation minister Julian Amery and French ambassador Geoffroy de Courcel signed an agreement consisting of seven paragraphs regarding the collaboration between both countries to build a supersonic airliner. Avery declared in the House of Commons on the day it was signed that the project had advanced much further than any other known design for an SST. The aircraft would have an abundance of opportunities "if we progress now to secure a substantial share of the world market for supersonic airliners. This is a chance that will not come again," the aviation minister said. In the agreement the basics of cooperation were defined as well as the equal sharing of all costs incurred during development and production and possible later revenues. It was also stated that both the medium haul and the long haul version would be developed, as well as a common organisation for aircraft and engine manufacturers being set up.

The UK was hoping that the agreement would also smoothen the country's path into the European Economic Community, better known as EEC or EC, but on January 13, 1963, this failed at the veto of French President Charles de Gaulle, who stressed: "Nothing will prevent a close relationship and direct cooperation, as both countries have already proven when they decided to build the supersonic aircraft Concorde together." At this time the name Concorde appeared publicly for the first time, which was originally suggested by the ten-year-old son of an employee of the British Aircraft Corporation (BAC). The British, however, at first insisted on spelling Concord without the "e" and it took until December 1967 for the name Concorde with "e" to finally be accepted by both sides. Beginning a formal collaboration between British and French was not easy, organisation and coordination between two management teams in different countries with rotating chairmen was slowing down decision-making processes. Also there was competition between the individual companies involved. "Nationalism was present on both sides," asserted British Concorde test pilot Brian Trubshaw in his memoirs

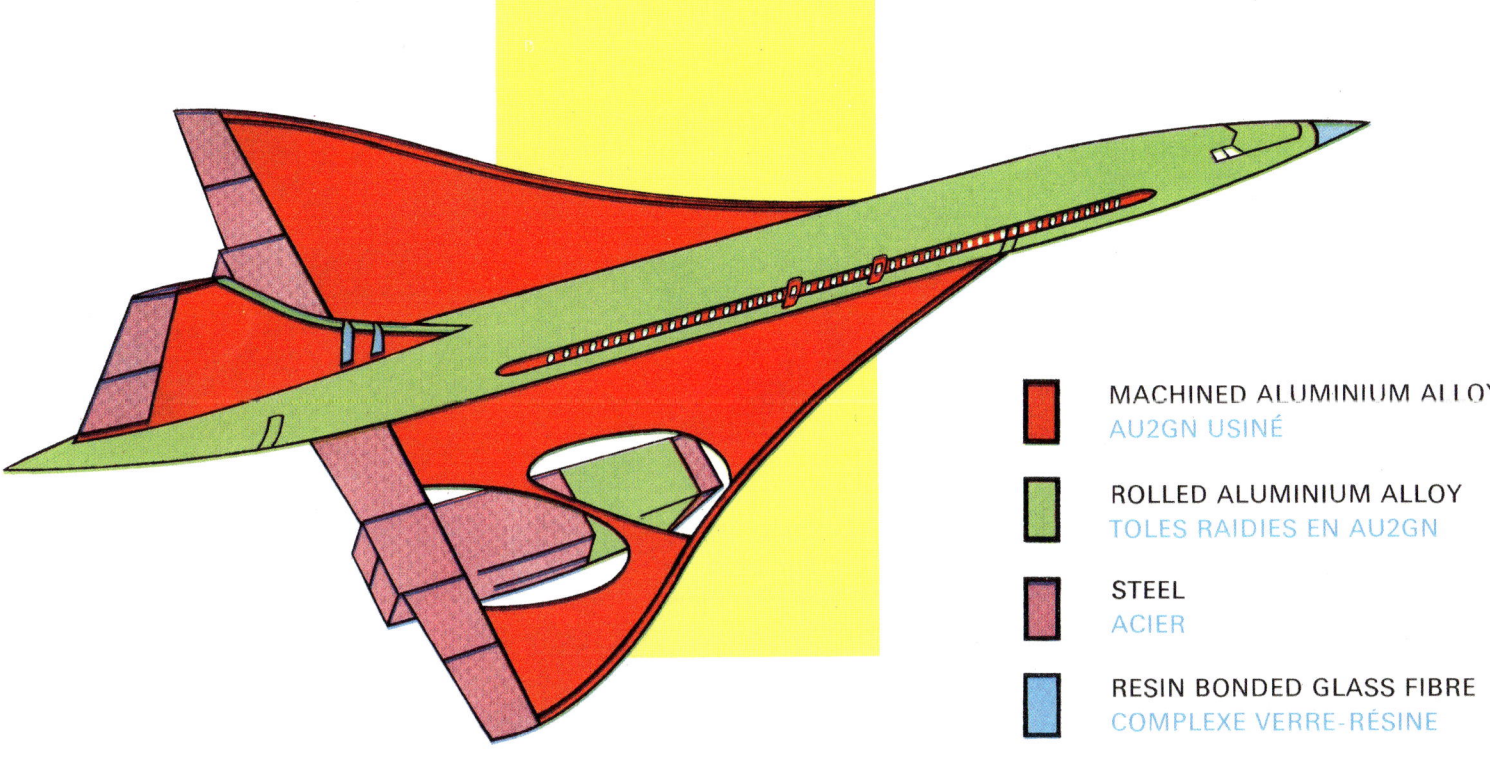

MACHINED ALUMINIUM ALLOY
AU2GN USINÉ

ROLLED ALUMINIUM ALLOY
TOLES RAIDIES EN AU2GN

STEEL
ACIER

RESIN BONDED GLASS FIBRE
COMPLEXE VERRE-RÉSINE

Most of the Concorde is made of aluminum alloys, only the flaps, rudders and engine cells are made of steel. [AS]

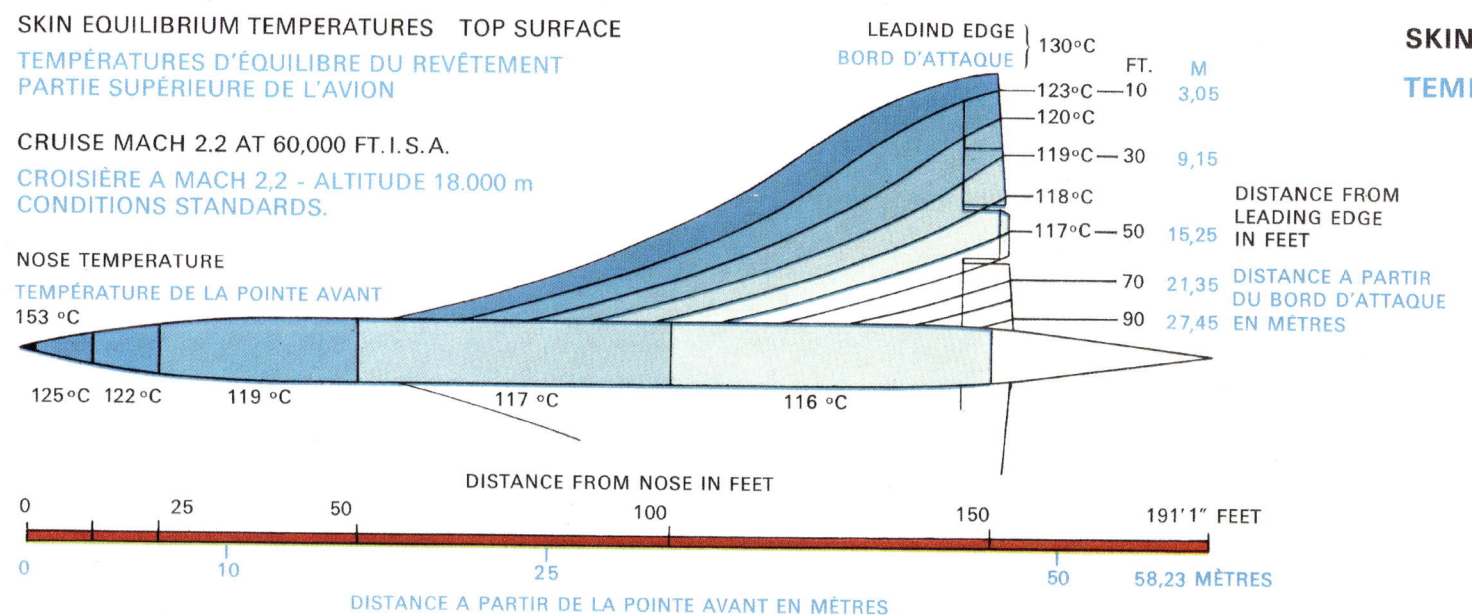

SKIN EQUILIBRIUM TEMPERATURES   TOP SURFACE
TEMPÉRATURES D'ÉQUILIBRE DU REVÊTEMENT
PARTIE SUPÉRIEURE DE L'AVION

CRUISE MACH 2.2 AT 60,000 FT. I.S.A.
CROISIÈRE A MACH 2,2 - ALTITUDE 18.000 m
CONDITIONS STANDARDS.

NOSE TEMPERATURE
TEMPÉRATURE DE LA POINTE AVANT
153 °C

125°C 122°C   119 °C   117 °C   116 °C

LEADING EDGE
BORD D'ATTAQUE } 130°C

SKIN

TEMP

FT.   M
123°C — 10   3,05
120°C
119°C — 30   9,15
118°C
117°C — 50   15,25

DISTANCE FROM
LEADING EDGE
IN FEET

70   21,35   DISTANCE A PARTIR
                 DU BORD D'ATTAQUE
90   27,45   EN MÉTRES

DISTANCE FROM NOSE IN FEET

0   25   50   100   150   191'1" FEET

0   10   25   50   58,23 MÉTRES

DISTANCE A PARTIR DE LA POINTE AVANT EN MÉTRES

At supersonic speeds parts of the airframe are exposed to high temperatures. This 1960s graphic shows the maximum expected heating. In reality the nose temperature as the hottest spot rose to a maximum of 127°C (AS)

Concorde Support Division | DESIGN AND MANUFACTURE

BAE Systems

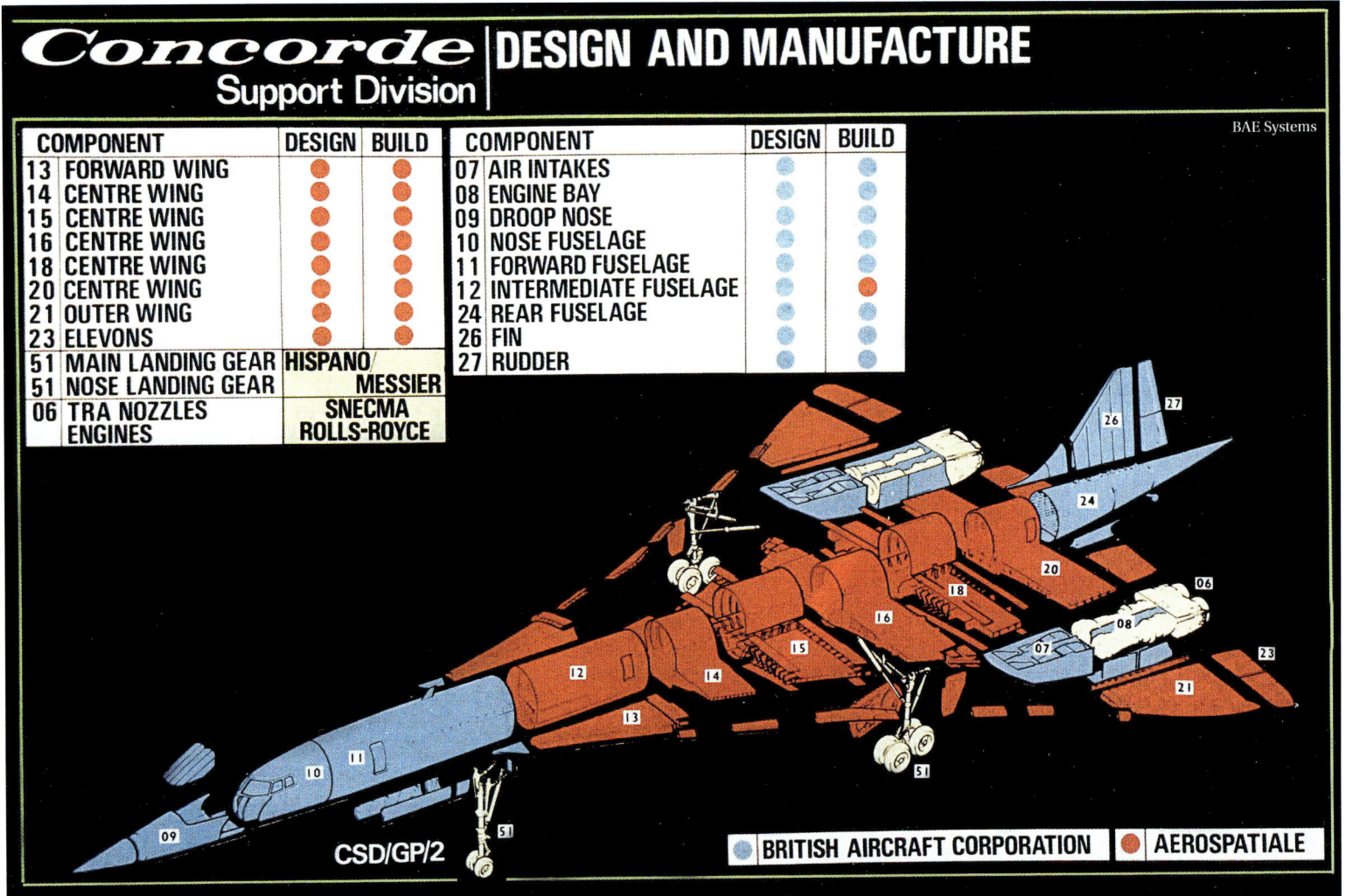

COMPONENT		DESIGN	BUILD	COMPONENT		DESIGN	BUILD
13	FORWARD WING	●	●	07	AIR INTAKES	●	●
14	CENTRE WING	●	●	08	ENGINE BAY	●	●
15	CENTRE WING	●	●	09	DROOP NOSE	●	●
16	CENTRE WING	●	●	10	NOSE FUSELAGE	●	●
18	CENTRE WING	●	●	11	FORWARD FUSELAGE	●	●
20	CENTRE WING	●	●	12	INTERMEDIATE FUSELAGE	●	●
21	OUTER WING	●	●	24	REAR FUSELAGE	●	●
23	ELEVONS	●	●	26	FIN	●	●
51	MAIN LANDING GEAR	HISPANO/		27	RUDDER	●	●
51	NOSE LANDING GEAR		MESSIER				
06	TRA NOZZLES		SNECMA				
	ENGINES		ROLLS-ROYCE				

CSD/GP/2

● BRITISH AIRCRAFT CORPORATION   **● AEROSPATIALE**

Production of Concorde was split between British Aircraft Corporation (BAC) and Aérospatiale, with one exception they each were responsible for design and manufacturing of their respective sections (BAE Systems Heritage)

benefits today. France took on 60% of the work on the fuselage, wings and structures, while the British had the leadership in engine development. In Weybridge the aft fuselage with the tail unit and rudders as well as the front section with the droop nose were assembled, while French partners built the landing gear, hydraulics, cockpit instruments and air conditioning. In principle, no work package was assigned twice, with the exception of the prestigious final assembly to be carried out both in Toulouse and Filton. "It is not unreasonable to look upon Concorde as a miracle," stated Brian Trubshaw, who was involved in every development stage and the entire test campaign and therefore uniquely qualified for such assessments. "Who would have predicted that the combination of two governments, two airframe companies, two engine companies – each with different cultures, language, and units of measurements – would have produced a technical achievement the size of the Concorde?"

At programme launch it was expected that both planned versions would come out at the same size with a hundred passengers each, both reaching Mach 2.2, with only the ranges being different, 4400 km for the medium haul version and 6000 km for the transatlantic model. The shorter haul Concorde

later, "each party harboured suspicions that that the other was trying to do them down". For Kenneth Owen, the positive prevailed: "It taught Britain and France how to design and construct the world's first in-service supersonic airliner. That is the achievement. It is unique, it is substantial. The result of 15 years' hard work was success, where it might have been disaster."

By 1964 the whole project was in danger as the new British Labour government was planning to exit the agreement. They had to realise, however, that this was hardly possible, according to the treaty signed in 1962 without paying heavy penalties. "When the Concorde was launched, Prime Minister Harold Macmillan had so little confidence that the French would keep it going that he insisted on a 'no-cancellation' clause," British politician Tony Benn recalled later. "When the Labour government was elected in 1964, it was that no-cancellation clause that prevented it from cancelling it as the Treasury wanted." Finally, all involved got their act together, and from the experience and infrastructure gained then, the European aircraft industry in the form of its successful Airbus partnership is still reaping the

was supposed to be equipped with air stairs underneath the fuselage and an air brake, while for the long haul version, a "limited afterburner system" was mulled. Both designs would not have provided forward sight for the pilots in cruise – a plan the engineers had to revise later. The principal companies involved in the programme were Sud Aviation (becoming Aérospatiale in 1970), British Aircraft Corporation (becoming a part of British Aerospace in 1977, today BAE Systems) as well as engine manufacturers SNECMA and Bristol Siddeley Engines (part of Rolls-Royce since 1966). While the French initially were under the assumption they were building two Concorde versions, for the British it was clear from the beginning that only a long-haul aircraft would make economical sense. It appeared to be more lucrative to fly London-New York in three and a half hours than Paris-Istanbul in 50 minutes. Only when Pan American signed options for six Concorde in the transatlantic version on June 4, 1963, did the French finally understood the British position. In May 1964, the last differences in the French and British designs were merged into one single concept, which now was aiming at carrying 118 passengers solely on intercontinental routes.

At the same time the competitive situation tightened dramatically as it became known that the Soviet challenger Tupolev Tu-144 was already under development. After nineteen months of preliminary procedures, serious development for Concorde began, besides the principal sites in Toulouse and Filton also in hundreds of other locations. In conceiving Concorde, expertise gained with supersonic fighter jets was only of limited use: military aircraft have to perform a wide spectrum of speeds from Mach 1.2 to Mach 2 and with angles of attack of up to 20°, but all only for a very limited time during a mission. Flights with supersonic speeds amount to only 12% of the total air time of fighter jets, while they fly subsonic for the rest of the time. A supersonic airliner, in contrast, has to perform continuous supersonic cruising for hours. Of the three and a half hours flight time between Europe and the US East coast, Concorde flew more than two hours at around Mach 2. "Concorde would travel at subsonic speeds only during takeoff and landing. This difference would later be illustrated when the tiny Concorde fleet would rack up more hours of supersonic flight during the life of the Concorde project than the sum total of all the military planes in the world combined," Samme Chittum asserted.

For Concorde, the intended speed was set at Mach 2 to 2.2. "That was a very wise decision," wrote test pilot Brian Trubshaw 30 years later. In subsonic cruise in the upper stratosphere, the temperature of the airframe was usually at minus 35°C, while this rose to plus 127°C at Mach 2 even in the thin air at high altitudes due to friction heat, and it would have doubled to about 250°C at Mach 3. To withstand such extreme temperatures one would have had to use steel or titanium as materials, but that would have created major production problems and big cost increases. Only in some specific places on Concorde that had to withstand high stress levels, were steel and titanium actually used. The British engineers working on Concorde had gained experience before in building the Comet, the world's first commercial passenger jet. And their French colleagues had brought the successful Caravelle to the world market

Assembly of forward fuselage sections at BAC in Filton, MSN 204 in the foreground, the first Concorde for British Airways, still BOAC at the time (BAE Systems Heritage)

two complete airframes had to endure static and dynamic loads. They were excessively heated up with infrared radiators and then rapidly cooled down again with liquid nitrogen, that's how thousands of flights were simulated. After concluding the tests, Concorde gained certification for 20,000 cycles (one cycle equals a takeoff and landing) – at the end of its active life, the most heavily used Concorde had accumulated just over a third of that. As a precaution, a commercial aircraft is normally only certified for half of the cycles that have been calculated as the maximum service life. In the case of Concorde, they took a step further and took only a third of the amount of cycles that were deemed possible in the tests as a certification limit. Initially, 6700 cycles were the defined lifespan of Concorde, later extended to 8500. With extra measures, this could have been stretched to 10,000 cycles, which would have enabled Concorde operations until the second decade of the new millennium.

Forward fuselage production at BAC in Filton, with MSN 202 in front, which later became the first British production test aircraft registered G-BBDG (BAE Systems Heritage)

Assembly hall at Aérospatiale in Toulouse, in the foreground the later F-BTSC for Air France that crashed in 2000 (Airbus Heritage)

that was even selling in the US. As revolutionary Concorde was as a technological challenge, the more down-to-earth and realistic the engineers remained in their ideas. As principal material a particularly heat-resistant aluminium alloy (hiduminium – AU2GN) was chosen, originally developed for engine fan blades. Today it is standard in aircraft manufacturing and used for example at Airbus. The exact list of materials Concorde is made of makes interesting reading: 35.2 tons per aircraft are aluminium, 15.6 tons steel, at least 4.7 tons are from titanium, 3.52 tons nickel, 2.16 tons copper, 1.32 tons chromium, 0.32 tons molybdenum, 0.09 tons manganese as well as traces of just 0.001 tons each of gold, silver and platinum.

Due to the constant change between very high and very low surface temperatures and a delay in heating and cooling of inner structural parts, significant thermal stress was created as an extra burden on the structure. Another problem was the cabin air cooling, for which the cooling capacity of the fuel was utilised. It first passed through a heat exchanger, which extracted the cold from the kerosene before it flowed into the engines. On approach, however, the fuel consumption of the engines wasn't sufficient for cooling the cabin, so parts of the heated fuel had to be pumped back into the tanks. In a specific and complex test programme

A crucial challenge was aerodynamics, as in supersonic cruise, the relation between drag and weight is extreme: every kilo of additional payload needs two kilos of extra fuel. The possible payload share of Concorde's total weight was one of the most important reference figures in the construction phase. In the production model it was set at 6% of takeoff weight, while other airliners achieve 12% to 15%. Half of Concorde's weight alone was used up for fuel, 31% for structure and 13% for propulsion. A priority was to minimise wave impedance, created by waves of compressed air initiating from the nose and engine inlets. The wave impedance stays smaller the sleeker the aircraft is, resulting in Concorde's fuselage shape resembling a lead pencil compared to the bulbous form of a Jumbo Jet. For Concorde the engineers chose delta wings with an ogival shape with a sweep of 55° and encompassing an area of 358 square metres. New insights about the flow ratios of delta wings had shown that they offer aerodynamic advantages in relation to wave impedance and allowed spreading the wing volume over a large area. Initially it appeared questionable for the designers if this wing shape, ideal for supersonic cruise, could deliver enough lift in slow flight phases at takeoff and landing

without compromising stability. The Europeans only considered conventional solutions from the beginning in contrast to the Americans, who for this very reason were intending to include swing-wings which then proved to be too heavy and complex.

With delta wings there are flow separations at the wing leading edges already at low angles of attack, but tests had shown that these separations resulted in stable flow conditions within a wide array of angles of attack with slender delta wings. For the first time in a civil aircraft, separated flows were taken as part of the normal flight conditions. There was fear that Concorde would need an extremely high angle of attack to attain sufficient lift. In contrast to usual aircraft configurations, Concorde did not have lift-enhancing elevators, for altitude and lateral control there were so-called elevons on the wings' trailing edges. But the concern was unfounded: it turned out that the vortex at the leading edges generated extra lift and also that a beneficial ground effect was enhancing lift close to the surface. Concorde needed much higher takeoff and landing speeds than conventional aircraft due to aerodynamic reasons, about 400 km/h at rotation and 300 km/h at touchdown. But then it made do with acceptable angles of attack of 12-14°,

In the former Concorde assembly hall in Toulouse (here the later F-BVFC under construction) was later used to build Airbus A300/A310 aircraft and today for ATR turboprop assembly (Airbus Heritage)

which still were considerably higher than for subsonic jets. The high speeds on the ground showed that Concorde was performing in a technological transition zone. Takeoff speeds of 400 km/h, limited by the maximum loads the tyres could take, were only achievable with extremely high engine thrust at normal runway lengths. The acceleration was double as much as in conventional aircraft, thrust had to reach 37% of the takeoff weight, while it is usually just 25%. Touching down at 300 km/h forced airlines to renew Concorde's tyres after less than 30 landings, while a Jumbo Jet touches down at 255 km/h and can use its tyres up to 200 times.

The wide speed spectrum of Concorde in flight caused a technical problem unknown at the time: the transition from subsonic to supersonic speed caused a significant shift of the centre of gravity, the aerodynamic centre of the aircraft, so that the force of the air attacks did not constantly remain in the same spot. While on takeoff the power of lift attacked at about 54% of the wings' depth on the fuselage, at a speed of Mach 1, it moved aft by about 20% of the wings' depth, while attacking at Mach 2 at about 62% of wing depth. This meant the aircraft had to be re-trimmed to prevent a tilting moment. A conventional, aerodynamic trim as on a subsonic aircraft would have created an unwantedly high trim resistance, due to the fairly measurable shifts of power. For Concorde, the problem was solved by transferring fuel from the main to the trim tanks in the wings as well as one in the aircraft's tail. At takeoff the centre of gravity was located at 54% of wing depth. If it had been only one per cent or 30 cm further forward, there would have been a requirement to considerably reduce takeoff weight, forcing the airline to leave 40 passengers behind. During acceleration to supersonic flight, kerosene was pumped from the main tank into the trim tanks in the aft wing and tail, shifting the centre of gravity by almost 5%. At the end of cruise, the tail trim tank was almost emptied and the remaining fuel pumped into the wings. To achieve the angle of attack necessary for landing, the system shifted the fuel in the aft wing area. The principle of tank trimming is still used in modern aircraft construction, as are many other Concorde innovations. As one of the first other aircraft types the Airbus A310 was equipped with a trim tank in the tail section.

The best aerodynamics would be inefficient if they were not coupled with efficient engines. For Concorde, the Olympus 593 was already chosen in 1964, a development from an engine conceived for the Avro Vulcan Mk1 bomber. It was supposed to deliver thrust to the planned supersonic TSR-2 fighter, which then was stopped on financial concerns, to the benefit of Concorde. The four Olympus 593 engines generated 69.04 tons of takeoff thrust, whilst during supersonic cruise 12.32 tons were sufficient. At an altitude of 53,000 feet (16,154 metres), 1.19 kg of fuel per kg of thrust per hour was burnt. Due to the wide spectrum of speeds there were was a wide array of requirements for Concorde's engine functions. Accordingly, the Olympus 593 was an entire system that also consisted of a variable-geometry air intake, the afterburner as well as thrust nozzles with variable exhaust geometry, all besides the core engine. The engine intake, with a length of 3.35 metres, had to slow down the incoming air during supersonic cruise from Mach 2 to Mach 0.5, and that without too many flow losses. A computerised movable ramp generated a system of shock waves, resulting in speed reduction. The air was already compressed before reaching the engine and heated up to 200°C. During takeoff, the ramps were fully upwards and the intake valves automatically opened to enable maximum air supply. As maximum power was required on takeoff, the engines themselves only delivered 82%, of which the intakes contributed 21% and the thrust nozzles 6%, the rest resulted from the inlet drag. In cruise, only 8% of thrust came from the engines, while 75% was generated from the intake.

Another innovation in the design of airliners, if very fuel-guzzling, was the afterburners or reheats. In the second

Fuselage production in Filton, in front MSN 201, later French production test aircraft F-WTSB (BAE Systems Heritage)

Concorde first flight in Toulouse on March 2nd, 1969 with André Turcat in command, Brian Trubshaw on right (BAE Systems Heritage)

combustion chamber, extra fuel was inserted into the hot exhaust coming from the engine and ignited – with Concorde's engines in this phase seemingly spitting fire. This system is usually just used in military airplanes, but to activate a significantly greater reservoir of energy. Turning on the afterburners provides a fighter jet with 50% extra thrust for brief periods, while for Concorde it was initially just 9%, which was later increased to 20% extra power on takeoff and 30% in the transonic zone. Activating the afterburners significantly increased fuel burn, so for the first nine per cent of an Atlantic crossing, 20% of the available fuel was used up. The convergent (narrowing)-divergent (widening) thrust nozzles were slightly shut on takeoff, lowering noise and enhancing performance. At higher speeds they increasingly opened up, they were an essential device for inflight power-control. The primary nozzles were closest to the turbine inside the engine cowling. On the outside of the very back of the engines were the clamshells or "buckets" as secondary nozzles, fully closed on landing to help braking with thrust reverse. Both inner clamshells could also be shut in the air to increase the sink rate.

Tight office: Concorde cockpit on a production aircraft

First flight of British Concorde G-BSST on April 9th, 1969 at Filton (BAE Systems Heritage)

Concorde boasted an innovation in flight controls that is a given in today's modern airliners, for example in the Airbus A320/A330/A350 families or Boeing's 777 and 787 competitors: fly-by-wire, meaning moving control surfaces such as flaps and rudders by hydraulic actuators initiated by an electronic signal, rather than moving them by a mechanic system with cables. Moving the pilot's yoke on Concorde sent an electronic signal to the hydraulics for the first time, but they were prudent enough to provide a mechanical backup system as redundancy in case of failure. Also the gear was moved by hydraulics, creating the longest "legs" in aviation history. The reason

for the gear's dimensions was the wing shape, causing the engines located near the trailing edges to come very close to the surface on landing due to the high angle of attack. To prevent scraping, the main gear needed enough ground clearance. As a final protection shield Concorde was equipped with a tail bumper wheel that only touched ground on an overly steep landing. A problem was posed by the long gear-legs after retraction, as they would haven taken up a lot of space. That is where a unique mechanism inserted into the shock absorbers came into play – it shortened the legs when retracted. Due to the high speeds on takeoff and landing, Concorde also needed highly efficient braking

systems. They utilised carbon brakes with antiskid, which has since become standard on other aircraft and many cars. Every wheel also had its own cooling system, as an aborted takeoff would heat the wheel rims up to 500°C, and even normal landings pushed their temperature up to 300°C.

One of the most visible characteristics of Concorde was its droop nose, and the separate visor in front of the actual cockpit windshield, retracting into the nose. Without this method, ground vision for the pilots on takeoff and landing would have been extremely limited, on approach, for example, a downward angle of view of just 5°. The nose, which had to be streamlined

for supersonic cruise, could be moved to four different positions on speeds slower than about 460 km/h. Three of these were standard on each flight, only the option to solely lower the visor was rarely used. For takeoff and when initiating final approach, the pilots lowered the nose by 5° and retracted the visor, allowing them a downward view of 10°. As soon as the gear was lowered prior to landing, the pilots lowered the nose to its maximum position of 12.5°, gracing Concorde with a birdlike outward appearance. Initially designers started with a nose that could be lowered by 17.5°, but the pilots didn't appreciate it, as they lacked any spatial reference when looking out of the cockpit windows. Test

pilot Brian Trubshaw found it gave "the feeling of looking over a precipice," while he liked the 12,5° nose lowering finally adopted, "a good thing because an improvement in directional stability resulted as well". Drag wasn't considerably increased when the nose came down, however, cockpit noise levels rose by so much that in these phases the pilots always used headphones even for their internal communication. The cockpit windows themselves were later also used in other aircraft and high-speed train locomotives. Whereas the passengers had to make do with windows not even as large as the palm of a hand, resembling peepholes. The prototypes were still equipped with larger

Acceleration of the British prototype G-BSST with ignited afterburners (BAE Systems Heritage)

Landing of the British prototype G-BSST after its first flight on April 9th, 1969 (BAE Systems Heritage)

Test flight of French prototype F-WTSS over the Pyrenees (BAE Systems Heritage)

windows, but the certification authorities opposed them. The danger of an explosive depressurization due to cracked windows at almost 20,000 metres of altitude seemed too menacing, as the cabin maintained an atmosphere resembling conditions at a more comfortable 3600 metres (modern aircraft such as the Airbus A350 have a cabin altitude at the equivalent of 1600 metres today, much more comfortable again). The experienced Concorde airline pilot and later Airbus test pilot Jean-Louis Chatelain summed it up very much to the point after the Concorde era ended: "If there is one word that best describes the Concorde, it is *advanced*. At the time it was designed, in the sixties, it was almost unbelievable that they dared to address such a challenging design. If you look at current aircraft technology, such as the Airbus technology, you can see it borrowed a great deal from the Concorde, which was a kind of laboratory for aircraft manufacturing in general."

It was a lengthy and sometimes arduous process from drawings to the first prototypes up until the first production aircraft. All technical enhancements to the serial version of Concorde were a result of the longest test phase in the history of civil aviation. It took from December 1967, when the first French prototype

Concorde prototype F-WTSA is seen here at cold soak tests at Fairbanks/Alaska in winter 1974 (BAE Heritage Systems)

001 F-WTSS was assembled, until the end of 1975, when Concorde received its official certification by the authorities permitting passenger service and the first aircraft were delivered to Air France and British Airways. In total, the flight test programme included eight aircraft (two prototypes, two pre-production models as well as four production aircraft), which altogether accumulated 5536 flight hours – three times as many as were necessary for the Boeing 747, which was also revolutionary. Even today, there are still big differences in how many hours of flight-testing are mandatory for entirely new aircraft, for the Boeing 787 it added up to almost 5000, while for the Airbus A350, about 2600 hours proved to be sufficient. During Concorde testing, the aircraft flew supersonic for a total of about 2000 hours, while the engines were tested in flight for about half of the total 52,000 hours of operating, with the remainder being ground runs. It was not making things easier that for political reasons many test facilities were installed in parallel in both countries. In a modern world, later asserted test pilot Brian Trubshaw, who had joined the Concorde programme at the end of 1965, all test flying would have been

Prototype F-WTSB taking off on a test flight from Toulouse (Airbus Heritage)

better than we would have expected from the simulator."

On April 9th, 1969, the British were finally ready as well. Prototype 002 G-BSST took off in Filton in front of the world's press and 12,000 employees of BAC and Rolls-Royce for its own first flight, also lasting 42 minutes. The flight under the command of Brian Trubshaw landed at Fairford, 80 km to the northeast, where the British conducted their test flights due to a longer runway. "My most important memory of 002's first flight was how uncomfortable it was for all of us to sit there with all the safety equipment on, consisting of helmets, parachute and all," recalled Jon Cochrane in 2003, then co-pilot. At the Paris Air Show in June 1969, both Concorde prototypes had their first major public appearance, next to the Boeing 747, also celebrating its world premiere. On October 1st, 1970, prototype 001 reached Mach 1 for the first time for nine minutes at an altitude of 10,970 metres. The honour to pass Mach 2 therefore fell to the British, who wanted to achieve

done in Toulouse, "but we were not that far advanced in collaboration to follow such a route," he recalled. At that time, the French were far more uncompromising, as they put up lavish facilities and hangars in Toulouse, which later gave Airbus a head start. They are still using some of them today for its successful production, as are regional aircraft manufacturer ATR.

The year 1968 passed with tests and preparations, in September the second prototype 002 G-BSST had been completed in England and started engine ground tests. Timelines got stretched and often turned out to be frustratingly lengthy for the engineers. This feeling was increasing when on December 31st, 1968 the competing Soviet SST Tu-144 took off for its successful first flight, while Concorde was still months away from a similar achievement. Almost four years to the day after the first metal had been cut, Concorde was ready for its first flight on March 1st, 1969, but could only perform it on March 2nd due to fog. French chief test pilot André Turcat commanded Concorde's first flight lasting 42 minutes, during which it accelerated to just 463 km/h, achieving a maximum altitude of 3050 metres. During flight, the nose remained in takeoff position, but the gear was retracted. Braking after landing was done with a braking parachute, which, unlike in the Tu-144 where it was a serial feature, was just installed in the Concorde prototypes. Turcat was pleased: "The aircraft flies

Prototype F-WTSS during brake tests in Toulouse in front of a barrier (BAE Systems Heritage)

All together at Fairford in 1971 – the three prototypes G-BSST, G-AXND and F-WTSS (BAE Systems Heritage)

this milestone on November 4th, 1970. But a fire warning in engine number two forced them to return home at Mach 1.35, while the French then made this landmark achievement on the same day, surpassing Mach 2 for 53 minutes at 15,300 metres of altitude. The first four Concorde aircraft were pure test vehicles hauling up to twelve tons of test equipment each in their cabins, enabling them to record up to 3000 parameters. The first two aircraft only resembled later serial models from a distance; due to insufficient payload capability and range they could not be deployed on the North Atlantic. Both pre-production aircraft that followed, F-WTSA

and G-AXDN, boasted fuselages stretched by 1.98 metres each, while the pressure bulkhead at the rear of the cabin had been shifted to the back by 4.78 metres. This enabled an increase of payload by two tons to 12.7 tons, the maximum passenger capacity of the pre-production aircraft being 139 passengers at a seat pitch of 34" (86 cm). For the many test flights, routes were needed that disturbed the population as little as possible, so the so-called Boom Alley was introduced. It stretched from the Irish Sea along the British West coast to the Bay of Biscay, there were also a few routes over the North Sea from northern Scotland to East

Anglia. For some supersonic flights, a straight route of almost 1300 km was required, under full radar coverage and easy to reach for rescue teams. For this task, a North-South route was chosen along the West coasts of both Scotland and Wales down to Cornwall.

As a result of the tests there were constant enhancements and improvements being introduced, for example on the autopilot, the wing leading edges or the engines. At the Hanover Air Show in Germany in May 1972 there was a special encounter: Concorde 002 was performing flights far away from its home base for the first time, but more importantly, Hanover

Cabins of test aircraft are rarely resembling later configurations in regular service, as seen here on prototype F-WTSB (AS)

British Airways Concorde G-BOAF makes an appearance at the Farnborough Air Show in the 1980s (BAE Systems Heritage)

also hosted the Tupolev Tu-144. "The arrival of the Tu-144 prototype just ahead of Concorde rather upset the apple cart by scraping all its exhaust nozzles on landing and frightening the air-show organisers to the extent that neither of us was allowed to fly," Brian Trubshaw remembered. "The length of the runway at Hanover was slightly short for the prototypes but Concorde had no problem stopping well within the distance available." Of course the aircraft test

crews were mostly interested in their respective competitor. "The Russian co-pilot spoke English very well with a strong American accent; sadly he was killed when the Tu-144 crashed at Paris," recalled Trubshaw. "We all got on extremely well, but found some elements of the Tu-144 combined aspects of a modern civil transport and a Second World War bomber." By June 1972 a multi-stop tour of the Middle and Far East started for 002, including Japan and Australia. In Tehran the Shah

of Iran joined for a ride, which was unexpectedly turbulent as the autopilot inadvertently sent the aircraft into a steep climb and then into a dive. "The Shah was well pleased and indicated that Iran Air would order three Concordes," recalled Brian Trubshaw. Such demonstration tours were normal tasks within the test campaign, the first flight overseas had already taken place on May 25th, 1971 when 001 landed in Dakar on its 142nd flight.

In December 1973, the first

serial aircraft F-WTSB took off in Toulouse and already reached Mach 1.57 on its premiere flight. "This showed that Concorde flight tests at this time were very mature," asserted Concorde veteran Brian Calvert. The serial aircraft were again shorter than the pre-production ones, boasted an all-glass visor and a longer tail cone, enhancing aerodynamics. In August 1974, the second production aircraft 202 G-BBDG flew to the Persian Gulf and to Singapore. "We flew along the Strait of Malacca and landed in Singapore at the end of an almost disappointingly short flight," reported Brian Calvert, co-pilot on this leg. Already early into the test programme, Concorde had shown a high degree of reliability, one of the British prototypes performed its entire month-long tour of Asia exactly on time as pre-planned for departures and arrivals. During the whole test phase there were no serious incidents.

An important part of the last test phase were the so-called route proving flights, performed jointly by the manufacturers and both airlines. Aérospatiale and Air France flew fully furnished Concorde 203 F-WTSC loaded with "passengers" (employees of both companies) and full inflight service on the planned routes to North and South America, while BAC and BA did the same with Concorde 204 G-BOAC on North Atlantic, Middle Eastern and Asia routes. For this

purpose, G-BOAC had already been awarded a preliminary type certificate allowing it to carry non-fare-paying passengers. The Olympus engines received their certification on September 25th, 1975, while on October 9th, 1975, French authorities in Paris finally awarded Concorde the *Certificat de Navigabilité*, its final type certification, the British *Aircraft Type Certificate* only followed on December 5th, 1975. The different times of awarding the certificates resulted from extra tests of the autopilot modifications that had been implemented after the experiences during route proving. Of the most elaborate test campaign in civil aviation history, 37% of flight hours were part of the development work, 18% of certification and 17% done during route proving and endurance testing. The remaining 28% were flown during demonstration, ferry or training flights.

Now Concorde was ready to go out and prove itself in scheduled service, although it was a given already that it would never recover its development costs and the taxpayers of Great Britain and France would have to pay the bill. But the overall balance remained positive, as Sir James Hamilton pointed out, a British engineer who was responsible for Concorde's final wing shape, becoming director general of Concorde on the British side from 1966. "Looking at the project as a whole, I think the design was basically right, and that's a considerable thing to say after more than 15 years since the design was started. There was a time when everyone was saying: 'You are designing a Mach 2.2 transport, you're building it out of light alloy, you're using a fixed-geometry wing. The Americans have got a variable-geometry aeroplane, built out of titanium, and going at Mach 3; they must be right.'

And they were wrong. They were wrong in every respect. … If we'd followed the American line the aeroplane would never have been built, just as their aeroplane was never built. Our greatest fear during the Concorde programme was not the American Mach 3 titanium aeroplane but that, at one stage (and Boeing got very near to it), they would build a bigger Concorde," summarised Sir James.

Concorde's wing profile is not only aesthetically pleasing but also more suitable it its aerodynamics compared to the Tu-144 (AS)

British Airways Concorde G-BOAC rests in 1977 at the airline's maintenance base in Heathrow (Stan Kingsley-Jones)

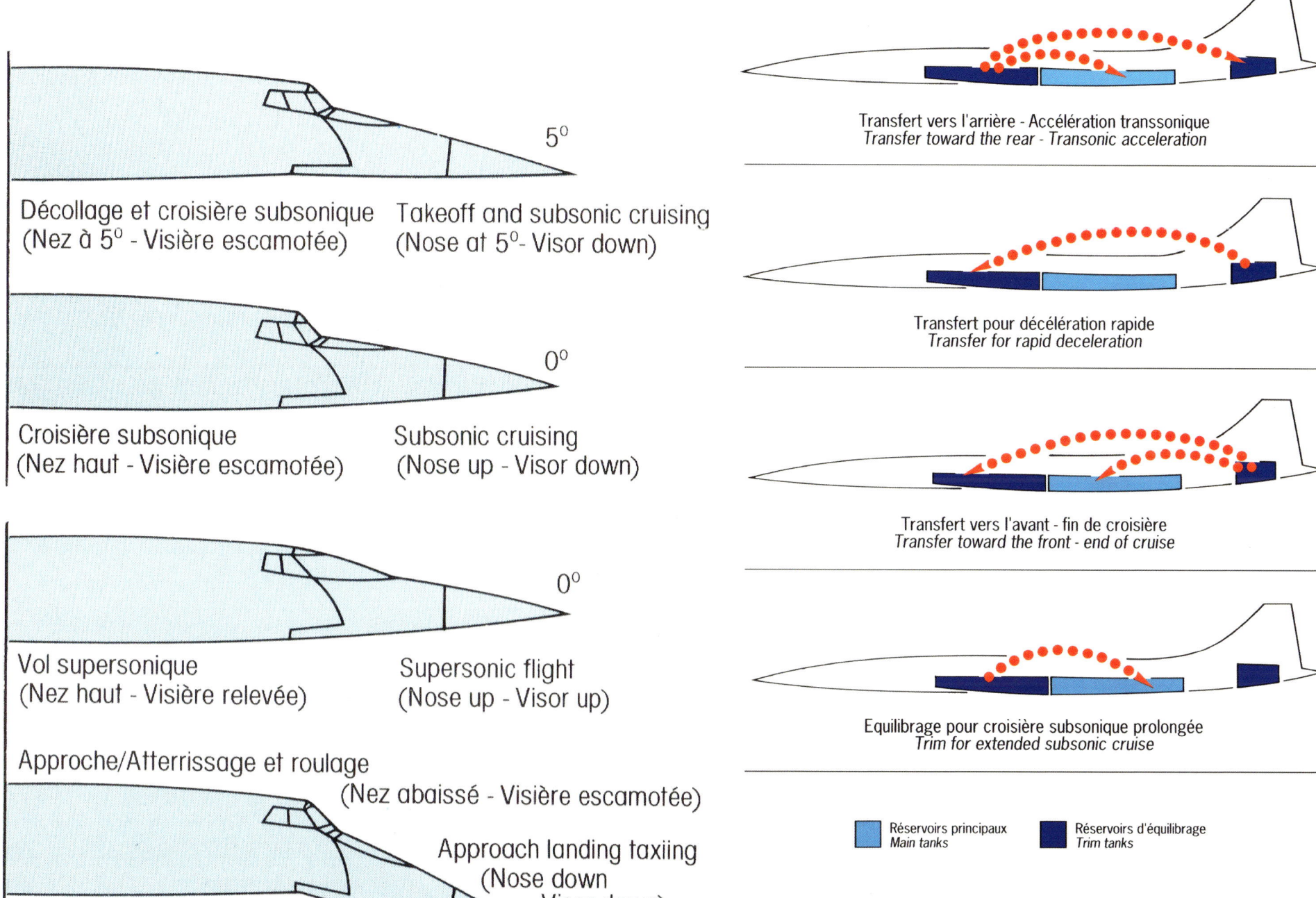

Décollage et croisière subsonique
(Nez à 5° - Visière escamotée)

Takeoff and subsonic cruising
(Nose at 5°- Visor down)

5°

Croisière subsonique
(Nez haut - Visière escamotée)

Subsonic cruising
(Nose up - Visor down)

0°

Vol supersonique
(Nez haut - Visière relevée)

Supersonic flight
(Nose up - Visor up)

0°

Approche/Atterrissage et roulage
(Nez abaissé - Visière escamotée)

Approach landing taxiing
(Nose down
- Visor down)

12° 30

Transfert vers l'arrière - Accélération transsonique
Transfer toward the rear - Transonic acceleration

Transfert pour décélération rapide
Transfer for rapid deceleration

Transfert vers l'avant - fin de croisière
Transfer toward the front - end of cruise

Equilibrage pour croisière subsonique prolongée
Trim for extended subsonic cruise

Réservoirs principaux
Main tanks

Réservoirs d'équilibrage
Trim tanks

TRANSFERT DE CARBURANT / FUEL TRANSFER

Nose and visor of Concorde allowed for four different configurations depending on the flight phase (AF)

Shifting the centre of gravity during flight was a decisive factor in flying Concorde, made possible by transferring fuel (AF)

seven

Concorde in scheduled service London and Paris to New York and more

"Concorde has given the European aircraft manufacturing industry a technical achievement … almost comparable with those earned in the United States and the Soviet Union by their space efforts."

Ronald Davies, 1998

British Airways Concorde in cruise (BA)

A rare sight, staged for a photo flight. Otherwise Concorde would never be seen in landing configuration above the clouds (AF)

The Tupolev Tu-144 once again had its sleek nose in front when it came to the inauguration of the world's first scheduled supersonic operation. At least formally, and that's probably exactly what the Soviets had calculated, always keen to gain prestige: they had beaten Concorde in first flying the Tu-144 already on December 31st, 1968, and the Russians did it again in starting "commercial" flights. On December 26th, 1975, Aeroflot began a so-called scheduled Tu-144 service between Moscow and Alma-Ata (today Almaty) in Kazakhstan, a distance of about 3000 km, albeit only to carry cargo and mail. At the time, the aircraft was still surrounded by secrecy and, according to all that was known in the West, incapable of operating regular passenger services. "So the news that the Tu-144 was in service had a hollow ring to it," former British Airways (BA) Concorde pilot Brian Calvert recalled. "It did not seem to have any bearing on what we were doing. Whether it yet was, or ever would be, an airliner had yet to be demonstrated." Just a few weeks later Brian Calvert accepted delivery of the first Concorde for BA and took G-BOAA from the factory airfield in Fairford to London-Heathrow. Air France had already received its first Concorde F-BVFA on December 19th, 1975, and a few days later on January 6th , 1976

they acquired F-BTSC, which was tragically lost decades later. After the longest test programme in the history of civil aviation, now everything was all set for the first scheduled supersonic passenger service.

However, the possible destinations were few, as Concorde was almost tailor-made to serve the high-demand markets of the US East coast, but there, of all places, it was not welcome at that time. In February 1975, the British and French had attempted to gain landing rights for New York and Washington DC. But the Americans were reluctant and pretended to act out of substantial environmental concerns, their main argument being the takeoff noise. Although it was clear both then and now that mostly prestige reasons and political considerations were behind it; the Americans could not just let the eye-catching Concorde slip in easily, after the USA themselves had so shamefully cancelled their own high-flying supersonic ambitions in 1971. In October 1974 the Europeans had already supplied the FAA with initial data, but the US authority even refused to permit Concorde to land for their route proving flights, test flights under real conditions, normally a well-established international procedure. Things dragged on, a complex environmental study had to be

prepared, in which the makers of Concorde mainly argued the six Concorde flights a day they were asking for would have a negligible influence due to the low number. The governments involved also made it clear that the technological progress and economical benefits it represented by far exceeded any possible environmental impact. They also stated that they would perceive it as a major discriminatory act should Concorde be denied this modest number of flights. But, no matter what, the US could not be considered as a destination for the inaugural flights. "The USA, most of all New York, which was supposed to be the most important market, made life difficult for us, not only because of noise," recalled Brian Calvert. "They told us to fly Concorde exactly like a subsonic jet and do holding patterns below 200 knots to remain within certain zones, but we wanted to fly 250 knots in the holding, and I flew holding patterns over Land's End for a full day to prove that we could still hold the aircraft in the allocated zone."

BA had only one destination available for the inaugural passenger flight, the tiny Kingdom of Bahrain in the Persian Gulf, half the size of London. "Bahrain was chosen for different reasons, most of all because the ruler accepted it," recalled Calvert. The Gulf state was to be an intermediate stop

Mach 2.05 on a test flight in the 1970s was faster than the usual cruise speed of Mach 2.02 (AF)

Braniff's domestic US Concorde flights between the East coast and Dallas didn't last much longer than a year in 1979/80 (AS)

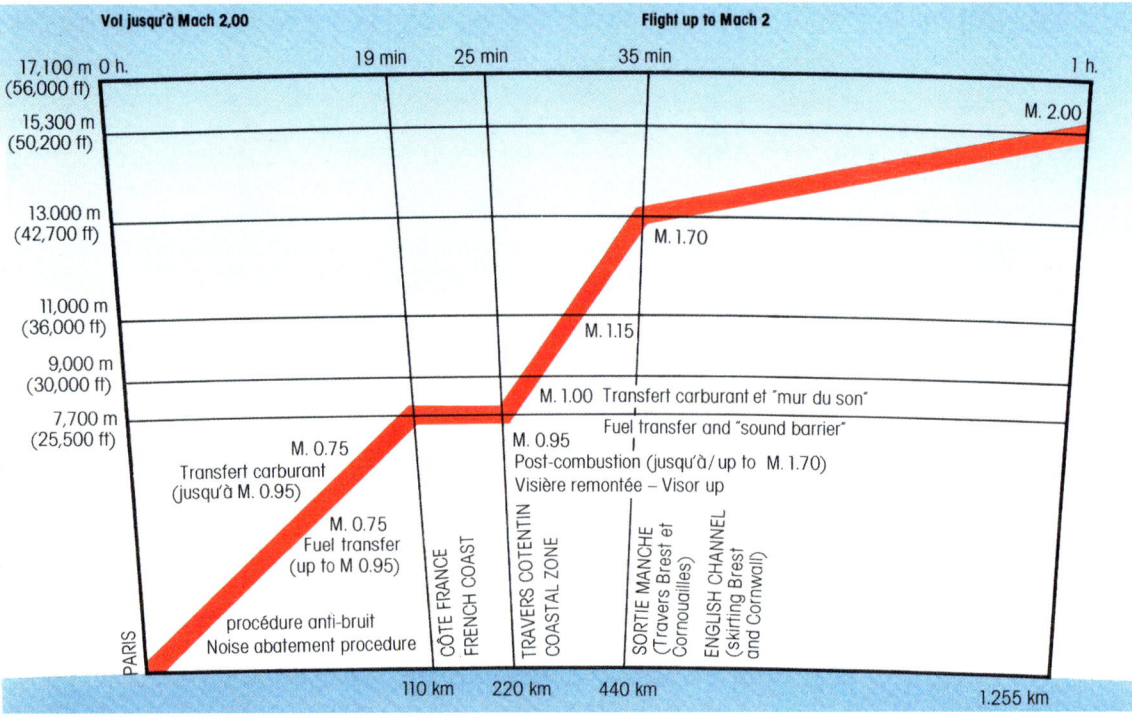

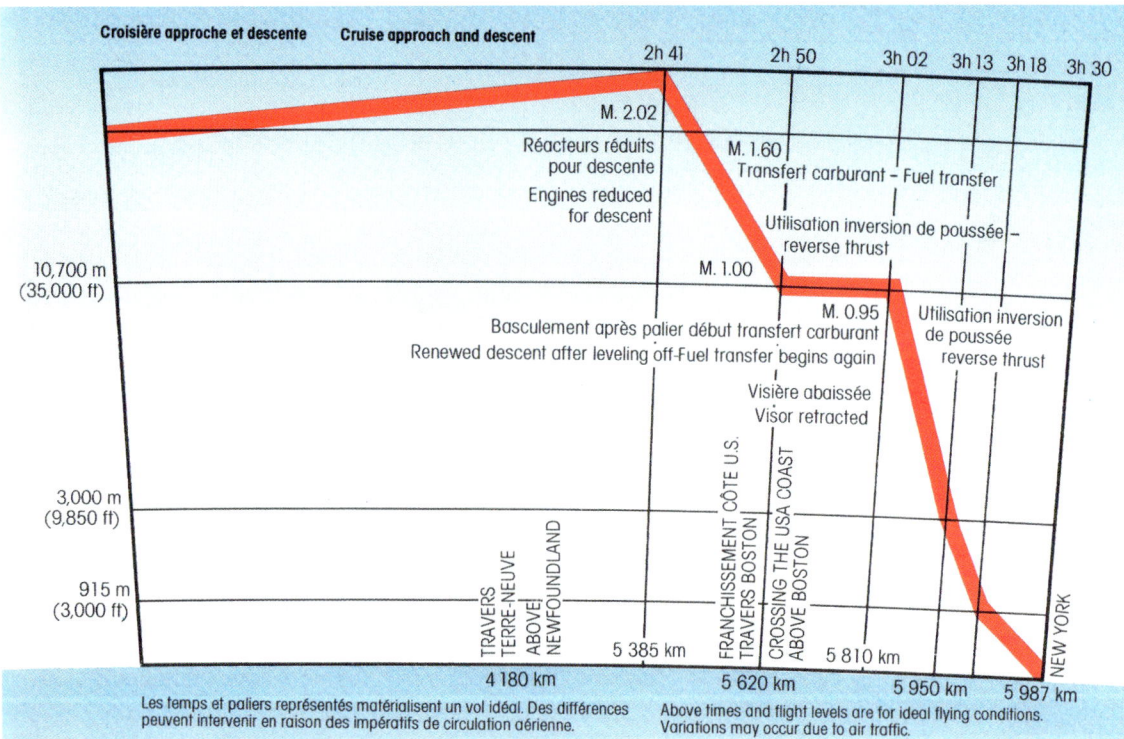

The profile of a typical Concorde flight from Paris to New York (AF)

in any case on the route via Singapore to Australia, planned for the medium term future. Though it was already obvious at the time that only non-stop routes would be economical, offering sufficient gains in travel time that passengers would be willing to pay a 20% surcharge on regular First Class fares. Also on this route a full return trip was possible in one day, making operations much more efficient for the airlines. On non-stop routes Concorde was cutting travel times in half, while the total elapsed travel time on longer routes was much less attractive. Destinations such as Singapore were not served non-stop from London either in the mid-1970s by aircraft such as the Vickers VC-10, the Boeing 707 or 747-100; this only changed when the Boeing 747-200 with greater range appeared. Air France was not deterred by an intermediate fuel stop on its inaugural flight, as it deployed Concorde from Paris via Dakar to Rio de Janeiro to usher in the age of supersonic passenger flying.

The big day was Wednesday, January 21st, 1976. "When I drove to work that important day I could feel the rush of adrenaline, we would write history today and everything had to work out," recalled Brian Calvert. At 11.40am Greenwich Mean Time, two Concordes began their parallel takeoff runs in synchronisation: G-BOAA in London-Heathrow on flight BA300 to Bahrain, and in Paris-Charles de Gaulle for flight AF085 via West Africa to Rio. A link telephone line was opened between both control towers to do a simultaneous 30-second-countdown, before both pilots were to release their brakes. The event was broadcast live globally and watched by an estimated quarter of a billion viewers, on a split screen they could see both aircraft at the same time. At Heathrow, 3000 people alone were packed on the roof of the now long-gone Queens Building, while all access roads were blocked by parked cars. On board the aircraft of both airlines was an illustrious mix of invited guests, led by the Duke of Kent on BA, and paying passengers, of whom some had booked as far back as 1969.

BA's flight was commanded by Norman Todd on the left seat, Air France's Concorde was flown by Captain Pierre Chanoine. The British aviation authority CAA had insisted at the last minute on sending two flight inspectors in the cockpit and it was agreed that test pilots Brian Trubshaw and John Cochrane would be declared as temporary inspectors for the purpose. "We were delighted to have them come along on the first services – they would be useful as well as good company," enthused Brian Calvert, First Officer on that day. Brian Trubshaw had a significantly different recollection: "One cannot say that our presence was exactly welcome… John Cochrane and I stood this farce for seven flights and then recommended to the CAA that BA should continue without us, which was accepted." Operationally the BA route to Bahrain was far more challenging than the Rio route of the French, which was mostly flying over the Atlantic. The BA flight, however, was proceeding subsonic at Mach 0.95, but still much faster than other aircraft, over France to Venice. After one hour "we ignited the reheat and climbed from 29,000 feet to 55,000 feet and accelerated to Mach 2," Calvert reported. In a long curve the aircraft passed over the Mediterranean Sea south of Crete and Cyprus, then north of Beirut over Syria and along the oil pipeline leading from Saudi Arabia to the Mediterranean, to the Persian Gulf. Due to these desert areas being very sparsely populated, Concorde was able to fly Mach 2 over land here. At 15.17 GMT, three hours and 37 minutes after releasing its brakes in Heathrow, G-BOAA was touching down in Bahrain. "We had done nothing extraordinary – had simply flown the aircraft from London to Bahrain, but we had written history in any case," remembered Brian Calvert. Passengers and crew were invited to a banquet that evening in the ruler's palace, who presented officials with a wristwatch each to

Rare Concorde group photo in flight shot at Christmas when there were no scheduled services (BA)

commemorate the occasion.

Meanwhile AF085 only needed a short subsonic stretch to reach the Bay of Biscay where it accelerated to Mach 2 over water. Along the Iberian Peninsula and leaving the Canary Islands behind, it flew to Dakar in Senegal, where the Concorde landed at 14.24 GMT. With a slight delay it was

airborne again at 15.45, when the crew was faced with the only technical glitch during the inaugurals. A spill door at one engine, which lets in a cooling air stream around the combustion chambers, was not opening, which delayed acceleration to Mach 1. But also this flight arrived at its destination at 19.05 GMT without major

problems, passengers and crew enjoyed a steamy party night at the Copacabana. A new era in aviation had begun, an occasion for which the Queen of England sent a congratulatory message to French President Giscard d'Estaing: "On the occasion of today's inaugural flight by Concorde aircraft of Air France and British Airways, I send

you and the French people my warmest congratulations. Today's flights mark the successful outcome of 14 years of close collaboration between our two nations. It is a source of pride that our countries have today inaugurated a new era in civil aviation."

When the festive mood had dissipated, the order of the

Concorde AIR FRANCE
FLUGPLAN
NORDAMERIKA

AF 001 tgl.	Concorde		AF 002 tgl.
11.00 ab	**PARIS** Charles de Gaulle	an	22.45
08.45 an	**NEW YORK** J. F. Kennedy	ab	13.00

* * *

AF 053 Mi · Fr · So	Concorde		AF 054 Mo · Do · Sa
20.00 ab	**PARIS** Charles de Gaulle	an	23.35
18.05 an 18.55* ab	**WASHINGTON** Dulles	ab an	13.45 12.55
20.50 an	**MEXICO-CITY** Internacional	ab	09.00*

*außer Fr   *außer Sa

SÜDAMERIKA

AF 201 Fr	Concorde		AF 200 Sa
18.00 ab	**PARIS** Charles de Gaulle	an	20.00
18.05 an 18.55 ab	**SANTA MARIA** (Techn. Zwischenlandung)	ab an	16.00 15.10
19.00 an	**CARACAS** Simon Bolivar	ab	09.00

* * *

AF 085 Mi · So	Concorde		AF 086 Mo · Do
13.00 ab	**PARIS** Charles de Gaulle	an	06.40
15.00 an 15.50 ab	**DAKAR** Yoff	ab an	02.40 01.50
16.00 an 16.30 ab	**RIO DE JANEIRO** Galeão **Boeing 737**	ab an	19.40 18.55
17.20 an	**SÃO PAULO** Congonhas	ab	18.05 Mi · So

gültig ab 1. November 1980

In 1980 there was a wide spectrum of Concorde services on Air France: From Paris to New York, Mexico City via Washington DC, Caracas via the Açores as well as Rio via Dakar (AF)

day was to manage day-to-day supersonic operations, a difficult task. BA for instance had only one aircraft, until in mid-February 1976 the second Concorde G-BOAC was modified for passenger services and ready to go. That's why the first Concorde timetable was quite modest: on Monday and Wednesday, there was a departure at 11am from Heathrow to Bahrain, returning the following day. These flights meant just about 16 flight hours for Concorde a week. BA hadn't expected this route to be particularly in demand, as Bahrain was just seen as a future fuel stop on the way to Singapore. But the load factor was at a surprising 48.5% for the first half year, about doubling predictions. Air France, in contrast, was already able to offer a "real" destination with Rio: Sunday and Wednesday at 1pm a Concorde took off from Charles de Gaulle airport to fly via Dakar to Rio, from where the return flight departed on the same evening at 8.30pm local time. This resulted in a weekly utilisation of about 25 flight hours. On April 9th, 1976, Air France started a new connection to Caracas in Venezuela via

Santa Maria in the Açores.

In the meantime, the battle for landing rights in the US continued. On February 4th, 1976, US transportation secretary William Coleman granted a 16-month test phase for Concorde flights to the USA after another lengthy hearing, with conditions attached: flights only between 7am and 10pm, adherence to noise-minimising approach and takeoff procedures, no supersonic flights above the US, and flights only out of London-Heathrow and Paris-CDG. A topic also discussed at the hearings was the environmental impact of Concorde, but the fact it emitted 2.3 kg carbon monoxide on 1600 km, while a car emitted

ten times that over the same distance, convinced even the sceptical Americans. The leading transatlantic carriers of the time, Pan Am and TWA, saw their premium passenger markets in danger at a time when they had already suffered from an economic downturn after the oil crisis. The US aircraft industry also saw its global dominance being threatened by Concorde.

But besides these perceived downsides, there were also advantages for America that finally turned the tide: trade and economic relations with Europe would be improved, the US would gain insights for possible future SST projects with its participation

As Concorde needed intense care between flights, so the aircraft, as seen at British Airways in Heathrow, spent most of their service lives inside maintenance hangars (BA)

For almost two years, the left side of G-BOAG appeared in the livery of Singapore Airlines in 1979/80, while the right side remained in BA colours (GT)

in analysing Concorde flights, and finally its permission would prevent possible retaliation by France and the UK. Without any Concorde having ever landed on American soil, fears about noise continued to play a major role. But it was accepted that Concorde would not be much noisier than the most widespread aircraft of the time, Boeing 707 and DC-8, while a maximum of eight flights a day were planned, versus over a thousand aircraft movements overall. Around the new Dulles

Airport in rural Virginia outside Washington DC, population was so sparse that only very few inhabitants were affected.

As the capital's new airport was run by the FAA itself things progressed rapidly. By on May 24th, 1976, both airlines sent their inaugural Concorde flights to Dulles airport. Pilots had to develop specific approach and departure procedures in the simulator before flights could proceed, to make sure all rules were adhered to, as several large zones of closed-

off military air space restricted the area off the coast. Initially BA was unsure about the operational consequences of these closures and therefore limited the maximum number of passengers into Washington DC to 80. First one to land on this sunny May day was G-BOAC from London, twelve minutes later the French touched down from Paris. Like in a film plot both aircraft moved on different taxiways in front of the control tower, where they came to a halt vis-

à-vis each other, lowering their noses and visors in parallel to greet America. The reception was mostly enthusiastic and load factors on the Washington flights were soon at 80%.

But the golden prize was flights to New York, and in that respect nothing much advanced. As, unfortunately, JFK airport is run by the Port Authority of New York and New Jersey, it doesn't automatically comply with a ruling of the federal government in Washington DC, a sensitive topic in a federal state. Many

neighbours around the airport still feared the noise and, lacking an actual experience of Concorde's noise levels, the Port Authority imposed a landing ban. As the airlines went to court, the authority was demanding a six-month noise study about the impact in Washington, Paris and London. "Bearing in mind that actual results from trials at Casablanca had been in their hands for months, one did not have to be a rocket scientist to smell the political game that was being

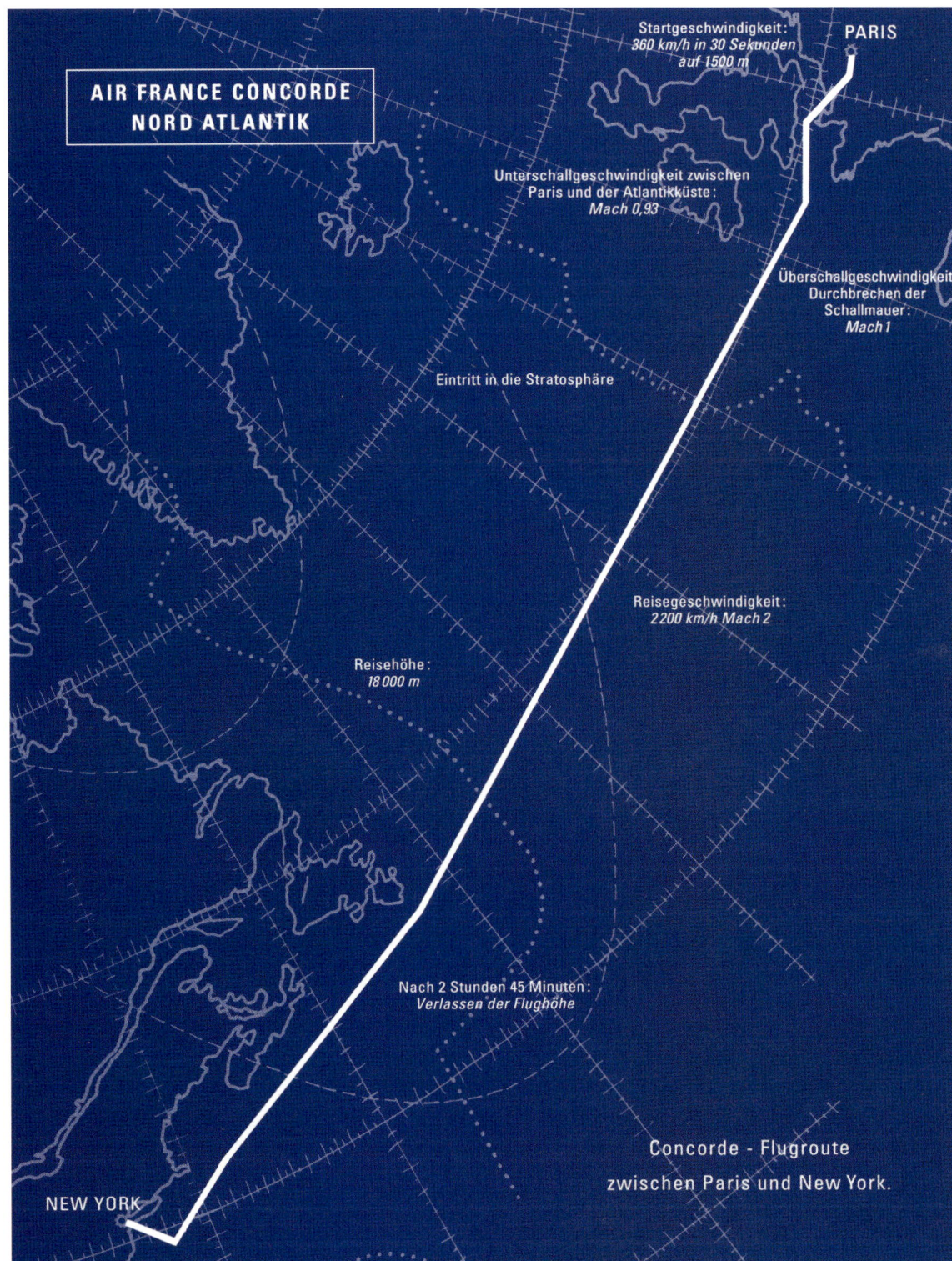

AIR FRANCE CONCORDE
NORD ATLANTIK

Startgeschwindigkeit:
360 km/h in 30 Sekunden
auf 1500 m

PARIS

Unterschallgeschwindigkeit zwischen
Paris und der Atlantikküste:
Mach 0,93

Überschallgeschwindigkeit,
Durchbrechen der
Schallmauer:
Mach 1

Eintritt in die Stratosphäre

Reisegeschwindigkeit:
2200 km/h Mach 2

Reisehöhe:
18 000 m

Nach 2 Stunden 45 Minuten:
Verlassen der Flughöhe

Concorde - Flugroute
zwischen Paris und New York.

NEW YORK

*For Concorde there were specific
transatlantic airways from Europe
to New York (AF)*

played," recalled test pilot Brian Trubshaw.

Once more airlines and manufacturers presented a detailed study about the noise footprint to be expected. They pointed out specifically that due to its superb manoeuvrability, it could even take off with tail wind in unfavourable wind conditions and thus avoid overflying populated areas altogether. Usually Concorde would not exceed the noise levels of the Boeing 707-320B, which in turn adhered to the permissible limits. But the authority didn't budge so court proceedings were set in motion. A lot of time had already been lost when judge Milton Pollock declared on May 11th, 1977, that the delay by the authorities and the ban of demonstration flights was "discriminatory, arbitrary and unreasonable" and lifted the ban. The authority appealed, there was some back and forth, but on October 17th, 1977, the US Supreme Court finally ruled in favour of Concorde. Now the airlines didn't want to lose even more time and acted swiftly – two days later a route proving flight was announced with French test aircraft 201 F-WTSB. The first Concorde at JFK came non-stop from Toulouse, approached over Jamaica Bay and landed

with a very reasonable noise measurement.

While the crew braced for a hostile reception, a lot of spectators had gathered and over 500 journalists peppered the pilots with questions. At takeoff the next day, the noise footprint remained modest, even though later fully loaded scheduled flights would be noisier. Especially on landing Concorde remained silent and often below the noise levels of other aircraft types. Now that the supersonic airliner showed up in reality it proved that many of the objections raised, based on theory, were unfounded, many of the protest groups that had taken to Manhattan's streets only shortly before now disbanded. Without doubt, however, the long uncertainty and delay of the New York entry had robbed Concorde of many market opportunities. "There was a heavy price to pay before this lengthy affair was over," assessed Brian Trubshaw, "several airlines cancelled their options because of having doubts they'd ever get permission for flights into and out of JFK." Brian Calvert had a similar reckoning: "Until Concorde actually arrived at New York, there was always a strong possibility that she would never be allowed to go there: having been and come back, she had won her right to the route for which she had been designed. There was never a final battle – only anti-climax when the enemy withdrew from the field."

On November 22nd, 1977, the longed for inaugural flights of both airlines into New York took place. Brian Walpole was in command of the BA Concorde, arriving 20 minutes early at the runway in Heathrow, as this time the Air France Concorde was supposed to land in JFK first. Over the PA he apologised to his premiere passengers: "I hope you forgive me for the short delay – after all, it's not much compared with the twenty months it took to get to this point today." In New York, anger and excitement had dissipated, after a short photo opportunity for photographers each aircraft taxied to its respective terminal, there was not even a proper press conference. Just the New York Chamber of Commerce hosting a banquet luncheon for about a thousand guests to celebrate the Big Apple's entrance into the supersonic era. But it showed quickly that the market had indeed been waiting for this offering, initially the Concorde flights enjoyed load factors of 80%. Depending on the state of the economy, there was more or less of a downward trend over the next years, at Air France the average loads fluctuated between 49.2% in 1992, to 66.2% two years later.

As soon as New York became a successful Concorde destination, BA dedicated itself to expanding the supersonic route network to Asia and on December 9th, 1977, started to fly from London via Bahrain to Singapore. But after only three round-trips this was stopped again, the Malaysian government objected to supersonic flying over the Strait of Malacca and lengthy negotiations followed. On September 20th, 1978, Air France opened the route from Paris via Washington DC to Mexico City, the elapsed travel time including the stop was only seven hours and forty minutes. As due to a lack of traffic rights no passengers could be taken on in the US to Mexico, the flight, later routed through New York, never became profitable and was finally ended on November 1st, 1982. An interesting experiment was the cooperation between Air France, British Airways and Braniff International Airlines from Dallas, a well-known carrier at the time. The idea was to connect the oil state of Texas to Europe in an appropriate way. So the Americans teamed up with both the British and the French in jointly operating Concorde between Dallas and the US East coast. This started on January 12th, 1979, and was operated subsonic over land with American registrations. On these flights Braniff supplied the cabin crews and the registrations were temporarily switched from G-BOAC for

Christmas gathering: Even though weather conditions were miserable, this group photo taken in Heathrow on a Concorde day off turned out iconic (BA)

instance to G-N81AC or N81AC, for the duration of flying within US air space the "G-" was simply patched over with a decal.

But even an oil-rich US state could not lower jet fuel prices, and as these soared, Braniff ended the collaboration on June 1st, 1980. By January 24th, 1979, the route from Bahrain to Singapore had been re-established and with a solution unique in Concorde history: G-BOAD was painted in the livery of Singapore Airlines (SIA) on the left side, while on the right it sported the usual BA colours.

No other airline besides SIA ever got this privilege – promotional photos claiming to show a Concorde in Braniff livery have been photo-shopped. Initially SIA had intended to lease a Concorde including pilots, but due to issues with the supply of spare parts and technical

2:52.59
Record Time: February 7, 1996

A big billboard in Manhattan touted Concorde's record to London in under three hours (AS)

service during the cold season from London to Barbados in the Caribbean, a sunny destination popular among the British high society with a clientele who spent money lavishly. The route was slightly longer than flights to the US East coast but, due to the usually stable favourable weather at destination, it required less reserve fuel for alternatives. If the weather situation was less predictable, Concorde would put in a fuel stop in Lisbon on its way from London to the Caribbean.

Finally, BA was left with only Washington DC thrice weekly (until this ended on November 1st, 1994), flights twice daily to JFK and in winter once weekly to Barbados as permanent Concorde destinations. In

addition there were charter flights, started in the mid-1980s, which made up about 10% of all flights by 2000. Air France, however, restricted itself to one daily flight to New York and conducted only a few charters. In a typical year, BA's Concordes accumulated short of 6000 flight hours, while Air France logged 2555 flight hours, not even half as many, resulting in an average of 425 hours flown by each French Concorde in a year. BA, in contrast, was able to bring every Concorde in the air for on average a thousand hours a year between 1978 and 1990, but that still resulted in less than two and a half flight hours per day for each aircraft. At the same time, every Boeing 767 in the airline had an average

A flight engineer in an Air France Concorde cockpit examines the route chart (AS)

personnel they went for this Solomon-like solution. But all these efforts were in vain, as demand on this route remained low. On average just about 40 passengers flew to Singapore on each flight whilst to break even, 70 to 80 would have been needed; the route amassed double-digit millions of losses a year. Decisive factors in this were the non-stop flights started between London and Singapore at the same time with the Boeing

747-200. On November 1st, 1980, this route also became a part of Concorde history. At the time, Air France accumulated massive losses on its Concorde routes as well, according to media reports.

It was hard to overlook: even before it could establish itself, Concorde had dropped out of the race on nearly all markets because it was incapable of flying long routes non-stop. On March 31st, 1992, Air France

ended its services to Caracas and Rio. BA, in contrast, extended its thrice-weekly flight to Washington DC on to Miami on March 27th, 1984, a suitable route for supersonic flights due to the coastal location of both cities. At least it lasted seven years, but still the end of it came on March 31st, 1991. In 1987, a new winter destination was inaugurated which was kept in the Concorde network until the very end: a once weekly

daily utilisation of almost 16 hours. BA's Concordes reached the peaks of their operational careers in the financial years 1979/80 and 1987/88, the only ones with more than 8000 hours of scheduled supersonic flying. So it was understandable that in 1982 plans were already being mulled of how to open up novel ways of utilising Concorde. Courier service Fed Ex for instance showed an interest in leasing three Concordes from BA starting in September 1983. This would have resulted in some design changes, for example in the pressure bulkhead, cabin floor or air conditioning. But the deal never materialised, as the British and French governments couldn't come to an agreement about it.

They had a major say about everything Concorde for a long time as it was the governments covering the operational losses for both state airlines and they had also shouldered all the cost for development and production of the aircraft so far. In 1978 alone, according to media reports, French taxpayers subsidised each Concorde ticket sold with about an extra 1000 dollars. Until this time, both countries had invested an estimated ten billion dollars in today's worth into the Concorde project. When the last five unsold aircraft (Numbers 203, 213 and 215 of French and 214/216 of British production) could not

Nosing in – Air France promotional shot taken in Paris (AF)

be sold despite all efforts, after Pan American had backed out, they were sold for "symbolic prices" (some sources say a thousand pounds each or even less) -instead of the list price of £23 million (£130 million today). So BA and Air France each had seven Concordes in operation, or rather had to accept them into their fleets.

Given this level of state support, all of a sudden BA was now able to reap profits with Concorde operations. After accumulating losses of £10.4 million between 1975 and 1980 (and Air France losing £36.8 million in the same period), the profitable financial year 1980/81 almost reversed all the earlier operational losses. BA spokesman Bill Stevens was quoted as such in July 1979: "We are in the lucky position that our government in fact gave us the aircraft. So we only have to bear operational costs." These mostly depended on fluctuating oil prices. However in 1984, "Iron Lady" Margaret Thatcher applied considerable pressure and put in front of BA two alternatives: either run Concorde under its own funding or cease supersonic

operations altogether. So BA purchased five aircraft from the government for £16.5 million (about £47 million today), while two others had been acquired for the symbolic price. BA also bought all other facilities for Concorde operations including spare parts and agreed to pay British Aerospace (today BAE Systems) and Rolls Royce for any replacement parts or further developments needed by itself. The French did not make comparable arrangements. BA's long-term Concorde general manager Brian Walpole later explained: "Never was Concorde closer to being stopped than in the negotiations between British Airways and government in 1982-84. ...The government expected us to stop and I believe that, if British Airways had stopped, Air France may well have been minded to follow. And that would have been the end of supersonic civil aviation for a mighty long time."

New York was the only Concorde destination served between 1977 and 2003 – while this Manhattan low pass is a montage (BA)

Supersonic Travel

For air-to-air shots like this one, company photographer Adrian Meredith had to use a Tornado fighter to keep track with Concorde in flight (BA)

eight

Getting up close and personal with Concorde in Heathrow on 9/11

Andreas Spaeth's personal account of visiting British Airways on the day America was attacked

G-BOAF after landing from its first post-accident test flight with passengers in Heathrow on 9/11/2001 (AS)

September 11th, 2001 was an exciting day for me even before anyone could know what this day was going to turn into. Since my first flight on Concorde in 1993 as a young aviation reporter, the supersonic airliner was close to my avgeek heart as the most incredible flying machine I have ever experienced, and that were quite a few before and after. I had extensively covered the Concorde crash in July 2000, and a German weekly paper had even bought me a ticket on a British Airways Concorde flight only days after the accident for an insane sum

of money (those were the days when media still had big travel expense accounts), just to have me report on how it feels to fly on it in the aftermath. That was before Concorde's general grounding on August 15th, 2000. So it was very exciting that I got invited to visit BA's Maintenance Base in Heathrow on September 11th, 2001 to get briefed by Concorde's chief pilot Mike Bannister and BA engineers on the technical progress made so far to ready Concorde for its return to scheduled service after over a year on the ground. The day was not picked by accident,

but to coincide with the first Concorde test flight carrying a full load of passengers after the grounding.

I had arrived from Hamburg a few hours early before the afternoon visit to BA. To kill time and enjoy the bustle of aircraft movements, I remember visiting the free, if slightly run-down rooftop terrace of the Queen's Building in Heathrow (demolished in 2009 to make room for the new Terminal 2). Hard to imagine nowadays, but access was even without any security check, climbing some dark and slightly filthy stairs, and there were no glass

walls as impenetrable physical barriers enclosing the terrace. Little did I, or anyone, know at this moment that these were the last hours of an open and easy-access rooftop-viewing terrace in Heathrow for a long time. And that all the aircraft I was watching taxiing towards takeoff for transatlantic flights this afternoon would soon be in dire straits, unable to complete their journey to North America, and either having to divert to alternates such as Gander in Newfoundland or return to their origin. I was thrilled every time I had a chance to do some spotting at a place like this.

Then I found my way over to the BA Maintenance facility, and while I was checking in, my wife called from home: Apparently a small aircraft had hit the World Trade Center in New York, which had happened at 1.46 pm BST, probably an accident, as it didn't appear to be a big airliner. Having visited the South Tower's stunning open "Top of the World" observation deck on the 107th floor often before myself, this apparent incident felt still like close to home and was bewildering. This was a time before smartphones or ubiquitous WiFi, we only had fairly primitive first-generation cell phones by today's standard, so all possible was communicating by phone calls or SMS text messages. Me and a few other attending journalists were taken to a meeting room, told to wait for Mike Bannister, who at that time was still on approach to Heathrow on his first test flight with a full Concorde cabin on G-BOAF. The "passengers" being BA staff that had won their tickets to ride in an internal lottery. When the journalists insisted, a TV set was moved into the room and we watched live coverage from New York. It was a horrific sight, with the North Tower of the World Trade Center up in flames on the top floors, the façade clearly smashed, but still it remained unclear initially what kind of aircraft had hit and why – as it was obviously a crystal

Before 9/11, Concorde had been grounded and being worked on in the hangars for over a year since the accident (Baz Glenister)

While events in New York unfolded, BA's Concorde chief pilot Mike Bannister (left) and engineer Claud Freeman presented the new Kevlar tank liners in Heathrow (AS)

York. Which was Concorde's main passenger market, after all, with many of its frequent fliers even having offices in the World Trade Center. Bannister took a short glimpse at the screen showing still one tower emitting billows of smoke, uttered "Oh, that looks horrific", then ordering with the authority of a Captain used to give commands: "Now we switch off the TV and turn to Concorde." This man clearly had his mission, and nothing would deflect his attention on what he was going to deliver to the journalists here and now as planned. It felt utterly weird even at this early point in the event unfolding. But of course we were here to get long-awaited first-hand insights into the complicated process of returning Concorde into service.

Minutes after Bannister had started his presentation, my wife sent an SMS: "Now a second aircraft has crashed into the other tower." That had happened at 2.02pm BST, and even without further information, this made it clear this was an orchestrated attack rather than an accident. It was surreal, and it was impossible to concentrate on what was said, even though I was hugely interested in the topic, normally. It seemed like I was the only one in the room at that moment who knew how horrifyingly things evolved in New York, how meaningless all that what was said had become

in comparison, while the whole outside world as we knew it seemed to crumble behind our backs. The screen remained dark. I'm sure after a few minutes, others here must have gotten messages as well, but no one dared to interrupt or leave the room, which today probably would happen instantly with smartphones transmitting real-time coverage, always and everywhere.

Bannister's presentation ended with him stressing how satisfied he was with the progress made and how much he looked forward to the imminent return of supersonic service. But did this even matter now? Wasn't the core transatlantic market for Concorde, bankers,

Working on Concorde's engine air intakes on 9/11 in readying it for service again (AS)

clear morning in the Big Apple.

At about that time, Mike Bannister entered the room with a big smile. I had met him before, even flew with him on Concorde in the cockpit jump seat sometimes. Obviously, he was in a kind of tunnel vision, so focused on Concorde and his just completed first test flight that he had not fully realized what seemed to unfold in New

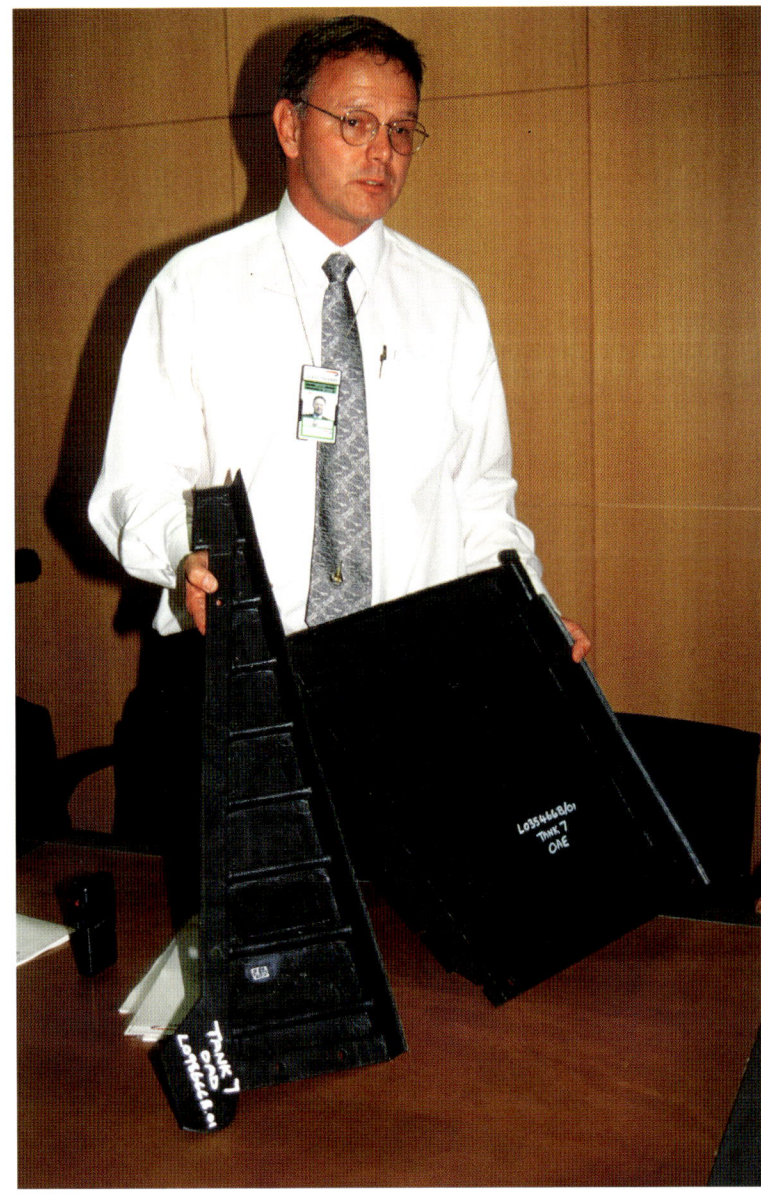

BA engineer Claud Freeman with the custom-made Kevlar tank liners in the afternoon of 9/11 (AS)

Author Andreas Spaeth flying Concorde after the re-start of services in November 2001 (AS)

happening again. While he was talking about the new radial tyres and the custom-sized Kevlar-rubber linings, 82 in total per aircraft, that been mounted on the bottom of all fuel tanks by up to forty mechanics working simultaneously on one aircraft, I received another SMS from my wife. Its content was even more shocking: "Now one tower has collapsed!", which had happened to the South Tower at 2.59pm BST. It was beyond comprehension, especially sitting in front of a dark TV screen, which could have let us witness this event of world-shattering consequences, talking about technical details of an aircraft instead which probably was rendered useless this very moment.

In my memory, we did not switch the TV on again, instead looked at tyre profile samples of the former Concorde tyre and the new radial tyre, custom-designed by Michelin to prevent big chunks of rubber coming off, it even if penetrated by a sharp-edged metal piece as in the accident. Then we went into the BA Concorde hangar looking at the real thing. Freeman even climbed into one of the fuel tanks, demonstrating the new Kevlar lining, smiling for photos. It was stunning and heartening to see the efforts being made to bring Concorde back, huge dedication and a lot of money making it happen. Still we all had by now switched

into a kind of tunnel vision, just ignoring what happened in real time right now at Concorde's main destination. Before we parted ways, Bannister took us outside to see G-BOAF parked in front of the hangar, on which he had just flown "passengers" a few hours before. It felt hollow somehow, and I did not know what to make of all of this, though I didn't even know at the time that also the North Tower of the World Trade Center had given way at 3.28pm BST.

But I had a feeling then that this meant the end for Concorde, one way or another. But not so soon, as it turned out. A little later than previously planned, only on November 7th, 2001 transatlantic supersonic services were restarted. But indeed, after 9/11, the already volatile market for Concorde never came back as it was,

speeding up the decision to retire it in 2003.

But on that fateful day in Heathrow, my first concern was if I would be getting home at all that evening. Overflying the City of London was prohibited, out of fear something similar to the US might be repeated, so there was a bottleneck of aircraft having to be re-routed. I watched some of the stunning and overwhelming footage from New York of which I had missed the live feed earlier at the BA Lounge. And also gave my first radio interviews from a cell phone ever that evening as an aviation expert, awaiting my delayed departure back to Hamburg that ultimately happened. Only two months later, I would find myself on my next Concorde flight to New York, there witnessing another catastrophe myself (see page 141).

brokers and other movers and shakers with endless expense accounts, collapsing right here and now?

After Bannister, Claud Freeman, BA's Concorde head

Engineer, was taking over, giving a superb presentation of how Concorde had been modified after the accident to alleviate concerns of the same sequence of events ever

G-BOAG being serviced in Heathrow after its first test flight with passengers in the afternoon of 9/11/2001 (AS)

nine

Concorde from London to New York in 2001

Between 1993 and 2003, Andreas Spaeth has flown eight times on Concorde. Here he details a trip on the supersonic airliner from London-Heathrow to New York in November 2001, shortly after Concorde services had resumed in the aftermath of the crash in July 2000 and the terror attacks of September 11th, 2001. Both events played a major role in the premature end of the Concorde era in 2003.

Note: The author travelled on the jump seat on another Concorde flight earlier, the descriptions and conversations from the cockpit given here actually took place earlier, as cockpit visits were impossible after 9/11. For illustrative purposes these have been merged into one text.

G-BOAC gets ready to fly to New York at Heathrow's terminal 4 in 2001 (AS)

On the way from New York to London with G-BOAB in 1995: The day flight Eastbound over the Atlantic reached London late in the evening (AS)

Flight BA001 leaves London's Heathrow airport daily at 10.30am GMT or UTC to cross the North Atlantic and fly to New York-JFK, covering a distance of 3150 nautical miles or 5833 km. The scheduled block time of Concorde on this route is four hours, arrival in New York (five hours backwards from the UK) is scheduled at 9.30am Eastern Standard Time (EST). That is why Concorde is nicknamed "time machine" as its speed enables it to arrive an hour ahead of its departure time, each being given in local time. Every departure is the result of a complex collaboration of very different contributors. About two hours prior to departure time, pilot, first officer and flight engineer gather for their crew briefing, everybody knows everybody among those who fly in the supersonic department. Concorde cockpit crews are a small elite club at the peak of their careers; Air France has a dozen cockpit crews while at British Airways (BA) there are 18 pilots, 15 first officers and 17 flight engineers. During the briefing the crew receives information about the weather en route and at destination as well as at possible alternate airports such as Shannon, Gander, Santa Maria in the Acores, Halifax or Washington DC.

In contrast to other transatlantic flight tracks, Concorde routes are relatively static, there is an Eastern, a Western and a backup route. Concordes coming from Paris and London in the morning are turning onto these special supersonic corridors as well as both aircraft leaving New York later in the day returning to Europe. According to schedule both aircraft take off with one hour between them, but if delays cause a closer separation, a Concorde might even have to wait on the ground to enable sufficient distance of 15 minutes of flight or 550 km from the other. After the briefing the cockpit crew puts together the flight plan and decides the central question – how much fuel to take. Criteria are the expected head or tail winds, temperatures at altitude during supersonic cruise (which can vary widely according to the phase of tropopause, with the general rule being: the colder, the more efficient the engines) and of course aircraft weight, depending on the current booking situation. On a typical flight, between 75 and 95 tons of kerosene are pumped into the tanks, but there always has to remain space for shifting fuel into the trim tanks to continually adjust the centre of gravity according to phase of flight.

An hour before departure, both pilots take their seats in the cockpit, much narrower than cockpits in any modern wide-body aircraft. Left leg first, head ducked, that's the only way for a pilot to take his or her seat, even though the seats can be moved far backwards on rails. Meanwhile the flight engineer has done his outside check, walking around the aircraft, today it's G-BOAF being readied for the flight: the first aircraft having been modified after the accident, and the one having been tested for the longest. In the cockpit, the crew are discussing takeoff procedures and determining important speed parameters depending on weather and aircraft weight: V1 (decision speed, after which takeoff cannot be aborted), VR (rotation speed) and V2 (safe minimum speed to climb even in the event of an engine failure). At the same time, the flight attendants (male and female, average age 32 at Air France, a tad older at BA) are gathering around their purser in the terminal for the cabin crew briefing, discussing the flight lying ahead: How many passengers are booked today? Are there customers with special needs or has anyone ordered a special menu? Which Concorde frequent fliers are on board? 80% of all Concorde passengers are repeat fliers, and a special service awaits them on Concorde – they are mostly addressed by name, and the crews even know their specific preferences. Air France even has a dedicated, well guarded

Concorde cockpit all lit up during maintenance (AS)

The promotional shot is deceiving about how cramped the cabin was. While the seats were almost at Economy level, serviced was First Class (AF)

for this preference. To buy entry into an exclusive club in which Madonna might be seated in the preceding row and Mick Jagger two rows behind, incognito and without bodyguards, that is the main draw even today for many Concorde customers to by a supersonic ticket.

In the meantime, passengers gather in the Concorde Room, a special lounge dedicated just to Concorde departures. Its walls are clad in elegant wood panelling and an array of armchairs by famous designers are standing by. In front of the panoramic windows the aircraft is prominently parked, while inside the upscale lounge hostesses are serving champagne or freshly squeezed juices, more recently there is even an entire á la carte breakfast menu available. Most guests are very relaxed, only a few are on the phone or hammering away on their laptops. Film star Hugh Grant lounges in a leather fauteuil with a slightly dishevelled hairdo and chats away on his cell phone. Later he gets up and holds small talk with a grey-haired businessman – Concorde frequent fliers casually know each other. At the same time another very familiar face casually sneaks into the lounge, hand in hand with his young wife Heather Mills: ex-Beatle Sir Paul McCartney, also one

card index stating exactly which mineral water a business man (80% of the guests are male) prefers, if a frequently flying opera star prefers meat or vegetarian dishes or which magazine a model reads. One of the most important questions at the cabin briefing therefore is: which VIPs do we have on board today?

Film stars, rock musicians, opera divas, politicians, royals or sport heroes love Concorde. On almost all flights there are some well-known faces in the cabin, as attested by the autograph albums BA cabin crews keep,

while Air France would never allow its cabin staff to ask customers for an autograph on board. Pop star Sting, for example, was such a dedicated Concorde lover that he insisted on going on board one of the test flights after modifications were done in 2001; BA refused even in his case. But, of course, he was on board the first flight after the forced break in November 2001 and, as the only VIP, made himself readily available to journalist's questions during and after the flight, he even assisted in the cabin pouring champagne.

On board Concorde there are a set of unwritten rules for prominent faces: anyone who is well known and wants to show his or her presence sits somewhere in the first three rows, which are specifically set aside for this purpose by the airlines until the last minute. Most coveted for wannabe-VIPs are the perceived best seats in the house in row 1, which are given away only just before departure. Anyone, by contrast, keen to be largely left alone and enjoy his or her quiet retreats to one of the last rows; Liza Minnelli or Michael Jackson are known

British Airways chief pilot Michael Bannister (left seat) on the way to New York (AS)

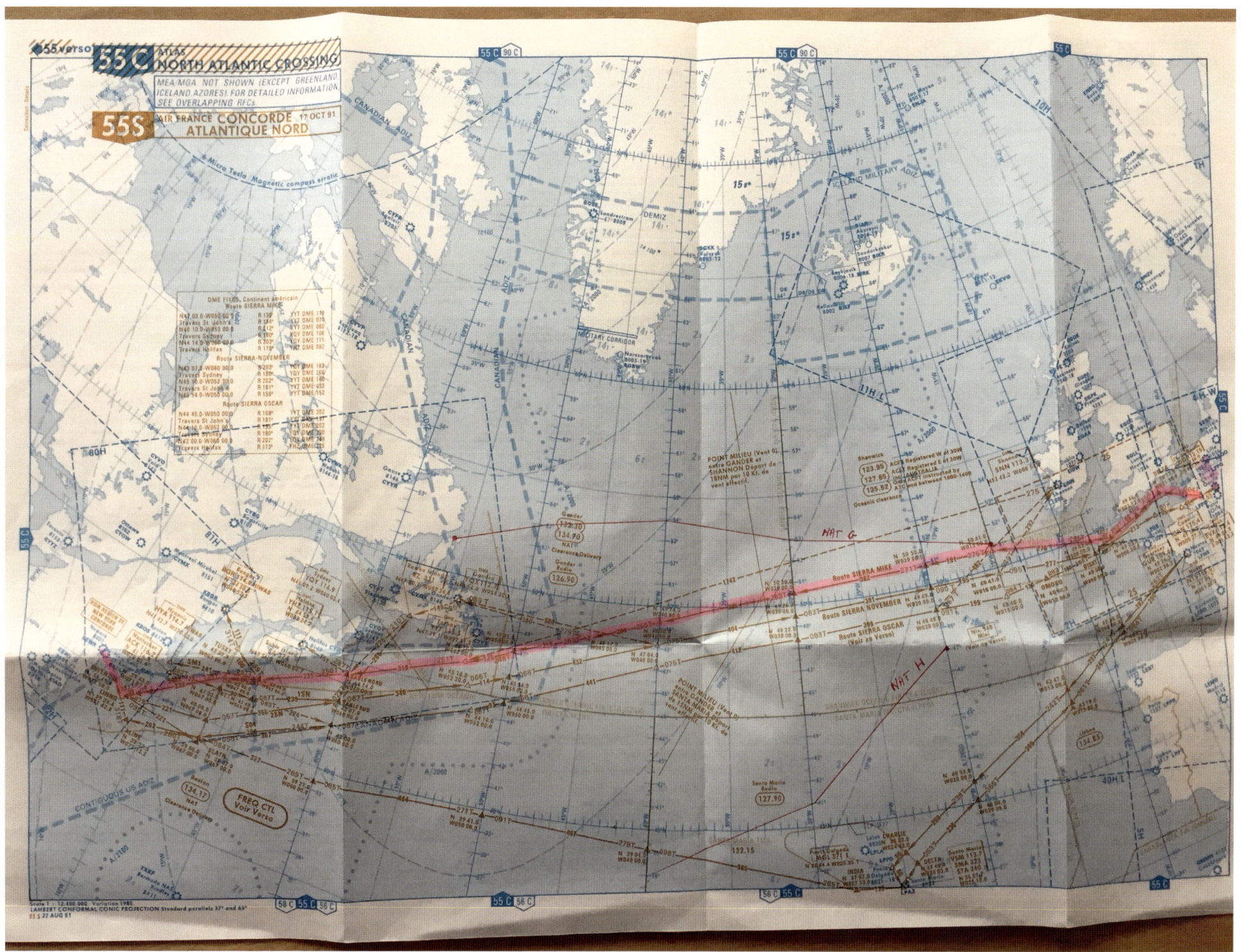

Paper flight charts were the norm during the Concorde era, here the Air France route from Paris to New York in July 1997 is highlighted (AS)

The complete catering offered on Air France Concorde flights between Paris and New York. Lobster and foie gras were available as well as three choices of mains (AF)

A bord du Concorde d'Air France entre Paris et New York - *On board the Air France Concorde between Paris and New York*

APÉRITIF

Caviar
Caviar

CHOIX DE HORS D'OEUVRE
CHOICE OF HORS D'OEUVRE

Homard et ses petits légumes
Lobster served with baby vegetables

Foie gras de canard et sa gelée de Porto
Duck foie gras in port aspic

CHOIX DE PLATS CHAUDS
CHOICE OF HOT DISHES

Méli-mélo de légumes cuisinés
Medley of vegetables

Turbans de sole à la tapenade
Turban of sole granished with black olive purée

Chateaubriand poêlé, purée de céleri
Panned tenderloin steack with celeri purée

Salade
Salad

Fromage
Cheese

DUO DE DESSERTS
DESSERTS DUET

Assiette de fruits frais Melon, mangue, fraises
Fresh fruit platter Melon, mango strawberries

Mignardises
Eclair au chocolat, tartelette fraise coco, macaron vanille

Petits fours
Chocolate eclair, strawberry and coconut tartlet, vanilla macaroon

narrow cabin, the passenger's coats are collected beforehand, brought aboard on a movable clothes rail and carefully stowed there. Some years ago, on an Air France flight, the ground staff forgot to deliver the coats to the aircraft before takeoff. Once in New York the swearing passengers stood in the snow without any gear, the vouchers handed out to buy new coats were only a small consolation to many.

Shortly before the doors are closed the pilots get the load-sheet handed into the cockpit containing the amount of fuel on board, the weight of the luggage, passenger number, takeoff and landing weights as well as calculations for the centre of gravity. The pilots sign it and the journey begins. At the very last minute a black

of the regulars on this quickest connection over the pond, and he is visibly pleased with the recently revamped lounge. When an elderly American in a wheel chair asks him for an autograph for his daughter, Paul rejects him politely but firm: "I never do this when I travel, but I can shake your hand". As it is time to board, everyone first goes through another security check of cabin baggage – unthinkable at least for Concorde passengers before 9/11. On entering the aircraft, almost everyone has to duck their head, as the doorframe is only 1.67 metres tall. This was preceded by another typical Concorde ritual: as there is very limited stowage inside the

The author's Concorde flights (indicated is the actual air time)	
27, January 1993	Air France F-BTSD CDG-JFK 3 hours, 27 minutes 20 seconds
14, December 1995	British Airways G-BOAB LHR-JFK 3:30:33
17. December 1995	British Airways G-BOAB JFK-LHR 3:17:49
18, July 1997	Air France F-BVFA CDG-JFK 3:31:26
28. July 2000	British Airways G-BOAC LHR-JFK 3:25:59
12, November 2001	British Airways G-BOAF LHR-JFK 3:18:39
24, Juny 2003	Air France F-BVFB CDG-Karlsruhe (FKB) 1:50:14
9, September 2003	British Airways G-BOAC LHR-JFK 3:22:28

Luxury in tight surroundings: Air France offered Persian caviar and foie gras on Concorde (AS)

limousine pulls up at the aircraft and delivers the last passenger, about which rumours were spreading in the lounge: Sarah Ferguson, the Duchess of York, better known as Fergie. She sits on seat 1A today. Right after she has buckled in, the aircraft is pushed back from the gate and the flight engineer does the engine-start up. Concorde chief pilot Mike Bannister welcomes passengers "on board the world's only supersonic airliner". In more detail than on other aircraft he explains the takeoff procedure for all those on board unfamiliar with Concorde. "For one minute and 16 seconds we will turn on the afterburners on

On every flight Concorde's fuselage was expanding by 20cm due to the friction heat. Result: A gap between instrument panel and cockpit wall (AS)

holding onto the aircraft from the outside. The afterburners have to be used sparingly as their use increases fuel burn enormously, by an extra 50%, while additional kerosene is injected into the hot engine exhaust. Soon nose and visor are drawn up again and immediately it becomes much quieter in the cockpit. By now, autopilot and auto-throttle have taken over and expedite the aircraft in 13 minutes to an altitude of 28,000 feet (about 9300 metres) and to reach the coast of the Bristol Channel. Up until here Concorde has to fly Mach 0.95, as supersonic flying is only allowed over water. After 17 minutes there is a sensation as if someone gave the aircraft a little kick in the back – the afterburners are active again. Just a minute later Concorde has passed the sound barrier, but nobody on board can hear it, only the Machmeter on the front cabin wall convinces the passengers. The aircraft is now flying absolutely stable and calmly, a clear sign that its aerodynamic features now work to its advantage.

As Mach 1.3 is reached an instrument on the flight engineer's panel shows how the engine inlets are moved downwards by computers to slow down and compress the oncoming airstream before it reaches the combustion chambers. The warmer the air

rotation and you might feel it when we turn them off again," he explains. Rotation speed in Heathrow this morning will be at 402 km/h, the aircraft being fuelled to capacity with 95 tons of kerosene.

"*Speedbird Concorde 001, cleared for take-off,*" comes the voice of the air traffic controller through the headphones of the cockpit crew. Captain Bannister pushes all four throttles as far forward as possible, and Concorde gets in motion right away. The difference from a normal jet can immediately be felt: The four Olympus engines plus afterburners delivering 20% extra thrust to make Concorde rotate despite its aerodynamic shortfalls in this phase, unleash such power that passengers are pushed much more firmly into their seats than usual. There is a smell of burnt carpet dissipating in the cockpit – "that's normal and always smells like this when reheat is on," explains the First Officer later. The droop nose is half lowered to the 5° position, the visor with the cockpit windshield is retracted. As Concorde accelerates, the First Officer calls out "V1!" before Michael Bannister pulls the yoke shortly and firmly towards himself and the aircraft rotates from the runway as predicted at 250 knots (402 km/h).

Immediately afterwards the reheat is switched off for noise abatement on the ground, on board one senses a slight delay, as if someone was briefly

After two hours at Mach 2, here cruising at 54,000 ft (16,700 meters), the nose was heated to 127°C, while here it is "only" 121°C (AS)

Machmeters in both cabins indicated the flight speed to passengers. Usually cruise was at Mach 2.02/2,190 km/h (AS)

is in this altitude, the longer the climb takes and the more fuel is burnt. On board there is no sense of speed, but a peek out of the tiny windows, just as big as a man's hand, show how Concorde performs now. Below it glides a Jumbo Jet that it easily leaves behind at Mach 1.54. The special feature of Concorde is that it cannot only be fast for a short period of time but continuously for up to two and a half hours, and without air-to-air refuelling. "Recently we were accompanied by a Tornado on a photo flight," explains Captain Bannister, "and we overtook him already at Mach 1.2 as it couldn't simultaneously accelerate and climb as we can."

The afterburners are deactivated again only nine minutes after their second ignition, as they are no longer needed above Mach 1.7. While in the cabin, today's 85 passengers get their second offering of the finest vintage champagne, the flight reaches Mach 2 exactly 38 minutes and 40 seconds after rotation in

Heathrow at an altitude of 49,000 feet (about 16,300 metres). Twice the speed of sound on this flight equals 2156 km/h. Fuel consumption in cruise is about 20.5 tons or around 25,600

litres per hour, and with tanks getting emptier and the aircraft lighter, it reaches up to 60,000 feet (20,000 metres) during the Atlantic crossing. At altitudes above 55,000 feet Concorde

pilots are allowed to climb without permission by ATC – as up here there is no other traffic anyway. This so-called cruise climb was practised by all aircraft in the early days of the

jet era, until airspace became too crowded.

Concorde flies in the upper stratosphere at the edge of space. That shows for anyone kneeling down next to a window

Vintage French champagne while cruising at the edge of space with twice the speed of sound was a standard on Concorde (AS)

seat in cruise and looking up: The sky is pitch black above the aircraft even in the middle of the day. Looking down, there is also often a rare view to be had: if the horizon is not obstructed by clouds or haze, one can clearly spot the slightly bent line of the horizon – proving nothing other than the curvature of the earth. "Only the astronauts in the space station ISS are now higher up than we are," says Captain Bannister, "they have to wear space suits while we sit here in shirts and move at 37 km a minute". But the aircraft seems to stand still, the upper edges of the next clouds are at least 6000 metres below Concorde, and there is a lack of reference points. Within the time it takes to read this short sentence, another three kilometres of distance have been covered.

The autopilot is now in the MAX CRUISE setting, the workload on the pilots low. Besides speed, the only limiting factor is the aircraft skin temperature. Although the air up here is extremely thin and the temperature very low, about minus 50°C, the friction heat is sufficient to significantly heat up the aircraft's aluminium alloy. The hottest spot is the tip of the nose, which reaches plus 127°C

after two hours of supersonic cruise, in the cockpit area the skin still measures 97 to 100°C, the wing leading edges 105°C while the aft of the fuselage stays coolest at 91°C. Even the windows in the cabin are clearly warmed up even on the inside, though every window actually consists of three layers. Exposed to heat all materials expand – and Concorde is no exception.

During every supersonic cruise the aircraft gains about 20 cm in length, and then on cooling down at slower speeds it contracts to the original length again; all cables on board have built-in flexibility that is needed, as well as all systems. This phenomenon can only be observed in one place on board: at the flight engineer's panel in the cockpit. Before takeoff it is attached tightly to the cockpit wall on its right, while after two hours of flight, a gap has miraculously opened up, one can easily stick a hand in. "If you leave it in there until landing, you'll be trapped," jokes Captain Bannister.

In the cabin, things are pretty busy at the same time, as the aisle is narrow and anyone wanting to go to the lavatory, which is equally tight, has to squeeze himself or herself through a narrow gap if the trolleys are out. It is hard work for cabin crews, as Concorde's angle of attack is fairly high in the on-going climb, so trolleys

have to be pushed uphill literally from the back to reach the front cabin. Despite the cramped quarters on board, Concorde service is decidedly individual, the flight attendants have some personal words for everyone and converse fluently in many languages. For frequent Concorde fliers trying to cut down on calories it might be welcome, for other guests craving a luxury experience it might be regrettable – BA has abandoned gourmet offerings such as caviar and foie gras on board Concorde some time ago. In contrast to its French competitor, which of course has a tendency towards French gourmet food, the British now serve rather rustic or hearty meals. Today's menu for example proposes a classic English breakfast with fatty sausages, eggs and fried bacon, as alternatives there are tagliatelle with wild mushrooms and asparagus, sea bass with Mediterranean vegetables or a salad with chicken and dried tomatoes. The wine list is exclusive, but there is not enough stowage capacity on board for more than a total of four wines (an exclusive Krug champagne of 1985 vintage, a red, a white and a port).

Also, the inflight entertainment is not up to the standards you'll find in First or Business Class of almost all airlines in subsonic aircraft, on Concorde there is just an audio programme with five channels and high-end headphones. The installation of video monitors at each seat would be possible, but according to market research it is not a priority for passengers. And indeed, most of them read or sleep after the meal, only a few are working. And there is no difference with today's VIPs: Fergie on seat 1A initially sorts material for her next children's book and signs photos, then takes a nap. Sir Paul McCartney behind in 2A enjoys his cheese plate, and just that, while Hugh Grant on 3C initially reads a well-worn pocket book and then engages in a chat with a white-haired manager next to him. Nobody has much space to spread out, seat pitch is 38" (96.4 cm), about the same as BA offers in its recently upgraded Economy Class World Traveller Plus on its Jumbo Jets. But still, the new Concorde seats on BA made out of blue Connolly leather are designed in such a way that they can be comfortably reclined and also offer footrests and adjustable headrests. And nobody, after all, has to sit on Concorde for longer than four hours, and there are no night flights.

Two hours and fifty minutes after taking off in London and at an altitude of 58,000 feet (19,300 metres), the pilots slowly retract the throttles and leave Mach 2 while almost unnoticeable starting to descend. The sink rate is about 5000 feet (almost

After landing in New York JFK Concorde had to blend into other taxying traffic with its lowered nose (AS)

1700 metres) a minute, the radar coverage soon shows Cape Cod on the US East coast and shortly after it comes into sight in reality on the right side. After three hours and one minute in the air, Concorde becomes slower than the speed of sound and now flies at Mach 0.95. "Even if we are still 15% faster than the other jets, we now have to follow the same procedures as them again," explains Michael Bannister. This shows as Concorde approaches the coast of Long Island at high speed when the air traffic controllers clear the approach into JFK. At this time Concorde has to adopt a higher angle of attack to produce enough lift. At an altitude of 2500 feet (830 metres) the angle is at 11°, which forces the nose to be drooped fully to 12.5° to enable sufficient ground vision from the cockpit.

By now the landing gear, including the small tail-bumper wheel, has been extended and locked while runway 4R in JFK is in sight. *"Speedbird Concorde zero zero one heavy, cleared to land four right wind zero seven zero at one five knots,"* the crew hears from ATC through their headphones. Landing speed is 300 km/h, the weight still 103 tons, immediately before touching down the aircraft is at its steepest angle of attack of 13°. "Concorde is very easy to fly on landing, good to manoeuver and very stable," Captain

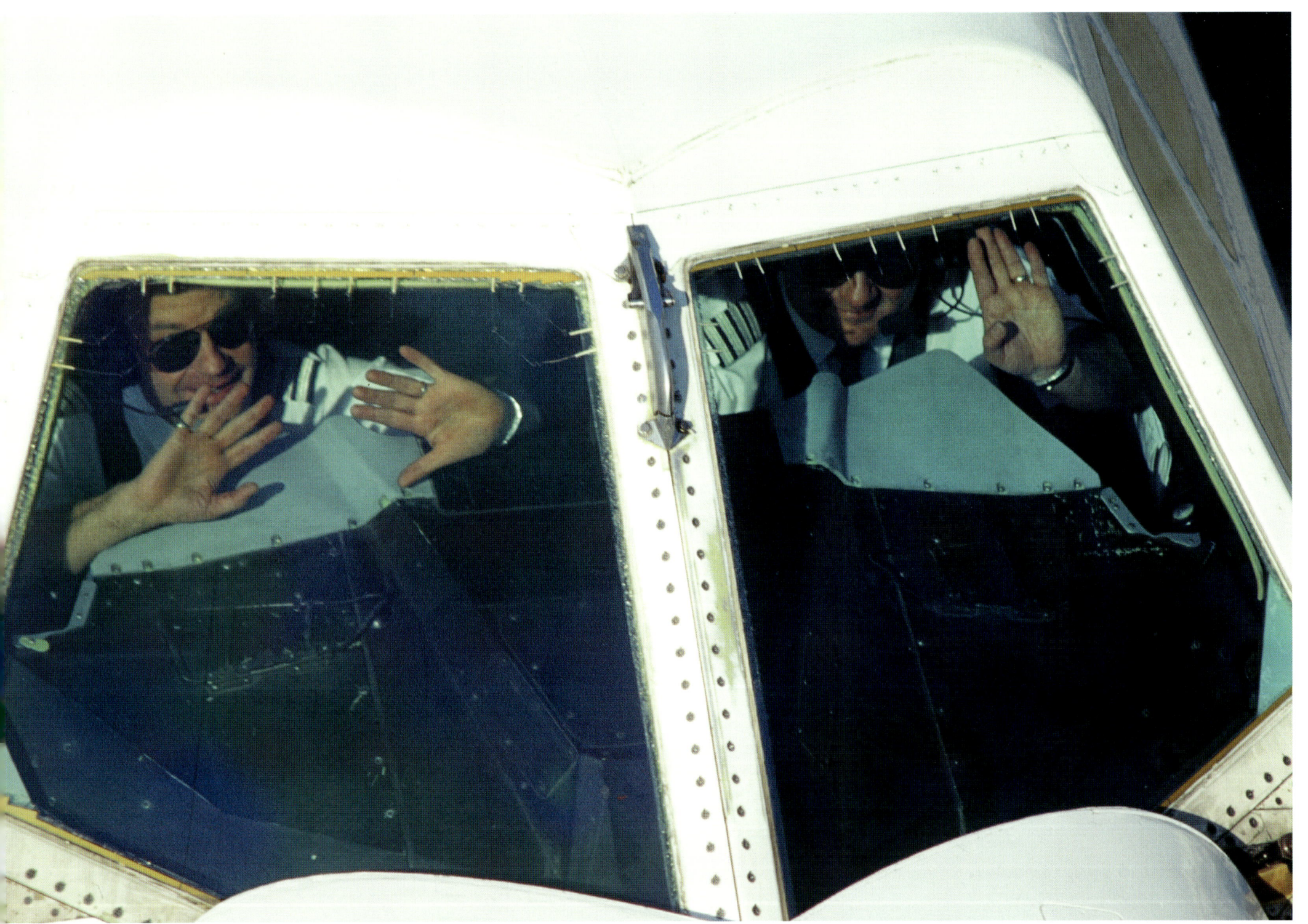

British Airways Concorde Chief pilot Michael Bannister waves from the cockpit at the return of Concorde flights in November 2001 (AS)

Bannister enthuses. From an altitude of about a hundred feet (about 33 metres) above ground the so-called ground-effect is experienced. The rapidly approaching vast wing compresses the air between aircraft and runway, producing extra lift as part of the engine intake air noise is also reflected. After exactly three hours, 18 minutes and 39 seconds in the air, the main gear of Concorde Alpha Foxtrot touches down in New York. The pilots immediately activate reverse thrust, seconds later as the front wheel comes down they briefly power up the engines again to lose speed through the reverse.

Right after leaving the runway the flight engineer turns off both inner engines, as at the end of a flight the two remaining ones are sufficient to taxi to the terminal. At this time BA's Concorde is the second aircraft arriving from Europe that day after the Air France Concorde, and with no one standing in line, entry formalities are finished within minutes. 85 passengers and nine crew members are now standing in New York, less than four hours after boarding in London. But the brain is hardly able to grasp how the body was able to move so quickly from one continent to the other. Transportation from the airport is now everyone's own responsibility; in earlier years, both airlines had offered free helicopter transfers with a Sikorsky S61 to 34th Street

The last Concorde interior on British Airways still appears timeless today, something one can't quite say about their 2001 uniforms (AS)

heliport in Manhattan, but then operator New York Helicopter ceased operations. Later there was a complimentary limousine service, all gone, reportedly due to a lack of demand.

This describes how things work out on an ordinary day and as the author has experienced several times. But the specific flight described in this chapter occurred on November 12, 2001, and that wasn't an ordinary day in New York City. Concorde had only resumed flying between Europe and New York after the accident in July 2000 and the terror attacks of September 11th, 2001 the week prior, on November 7th. Tensions were still high in the US, understandably. At 9.15am local time that morning, only minutes before Concorde Alpha Foxtrot was on approach from London, a horrific aircraft crash had happened in nearby Queens. An American Airlines Airbus A300 had just taken off with 260 occupants on its way from JFK to Santo Domingo when it hit wake turbulence of a preceding Boeing 747. To counter the turbulence, the First Officer moved the rudder so hard that the vertical stabilizer separated entirely and fell into the waters of Jamaica Bay. Seconds later, the aircraft crashed into a neighbourhood on the Rockaway Peninsula of Queens, all 260 souls on board perished. The black plume of smoke above the crash site was observed by passengers, including the author, from Concorde on final approach. Their aircraft was the last allowed to land for most of the day, as New York went into attack mode again, fearing another terrorist plot. JFK airport closed for about nine hours, FBI agents patrolled the terminal buildings, all tunnels to Manhattan were closed off, several landmark buildings such as the United Nations and the Empire State Building were evacuated because of terrorist attack fears.

As Concorde passengers stepped off their aircraft, they did not know what was happening - smartphones didn't exist at the time. Initially, even immigration was closed down, leading to the illustrious bunch of passengers having to wait and stay in front of the counters, with Sir Paul McCartney handing out chewing gum to some of his fellow travellers, all trapped for some time before immigration finally processed them. It was a very unsettling experience for everyone.

Despite the tragic events, it came as a bit of relief later in the day to all that this had been "just" an accident, though the second-deadliest disaster in US aviation history, but was not another terror attack. The author managed to catch a subsonic flight back to London that evening, having been trapped at a closed-down JFK airport for the whole day.

The last Concorde cabin for British Airways was the work of Factory design with Sir Terence Conran, the seats upholstered in Connolly leather (BA)

You have reached your destination – Concorde at gate of the British Airways terminal at JFK after arriving on the morning flight from London (AS)

Concorde from London to New York in 2001

ten

Crash, restart and the end of the Concorde era 2000-2003

"Concorde was a super-modern anachronism."

Wolfgang Tillmans, photographer, 2017

An iconic snapshot of world history: The doomed Concorde after taking off, trailing flames. A passenger on another flight nearby took it with a disposable camera

The crash site in Gonesse next to Paris CDG airport (French interior ministry) (c) MINISTERE DE L'INTERIEUR, FR. 2000

TEAM BUSH TAKES AIM

Newsweek

THE INTERNATIONAL NEWSMAGAZINE

August 7, 2000

After the Crash

The Concorde: What Went Wrong

The Future of Supersonic Travel

July 25, 2000: AF 4590 shortly after takeoff

US-magazine Newsweek puts the spectacular passenger photo on its cover (AS)

Paris Match puts a shot by Hungarian airplane spotters on the cover (AS)

TRAGEDY

The first CONCORDE crash claims 114 lives, and clouds the future of the "big white bird"

American magazine Time opts for a split title page (AS)

In the summer of 2000, Concorde had been commercially flying for almost 25 years without any hull loss or anyone being injured let alone killed during its operation. No other airliner can attest to an equally unblemished record over such a long period, even if other types were flying in much larger numbers than the 13 Concordes ever used commercially. Given the extremes it had to withstand in supersonic flying, this was a remarkable achievement. Especially given the fact that the Russian counterpart Tupolev Tu-144 suffered two hull losses in its much shorter active period, claiming a total of 16 lives. Time and again there had been incidents in Concorde operations, but as there were not any victims or even a crash, the French especially were keen to keep information about them under wraps. In the UK and France, Concorde was a national status symbol, to criticise which was deemed as unpatriotic. The former CEO of British Airways (BA), Baron Marshall of Knightsbridge, summed it up as follows in 2009: "As the embodiment of far-sighted vision, technological supremacy and sheer human endeavour, Concorde won a special place in the hearts of people right across the globe, but especially here in Britain where the aircraft had become a national icon. ... This precious piece of equipment was much more than just an aircraft." Even justified technical objections were put forward only by the US National Transportation Safety Board (NTSB), if at all; never were any doubts recorded by authorities of the two manufacturing countries.

For 96 Germans, two Danes and one each from the US and Austria, it was to be the trip of a lifetime that was supposed to begin on July 25th, 2000, in Terminal 2 of Paris Charles de Gaulle airport. They first met in the Concorde lounge. The predominantly well-off Best Agers had booked a 15-day cruise on the MS "Deutschland", which was already awaiting its passengers in the Port of New York. From New York City through the Caribbean and the Panama Canal to Ecuador was the ship's planned route, and for today's equivalent of a £1750 / US$ 2450 / 2000 surcharge even the outbound flight in a specially chartered Air France Concorde was included. Charter flight AF4590 was scheduled to depart at 15.30 hours local time to New York JFK. At the gate, Concorde F-BTSC was standing by, the third production aircraft having had its first flight on January 31st, 1975. At the end of July 2000, its logbook showed 11,989 flight hours. Not a critical age for a well-maintained aircraft and very few hours for 25 years of operations.

But departure of the cruise passengers was delayed, initially as baggage from the transfer flights from Germany was missing. Then, experienced Captain Christian Marty (54), having accumulated almost 13,500 flight hours in 34 years of flying, ordered a faulty thrust reverser on engine number two to be replaced. A source for many speculations later, but engine manufacturers Rolls-Royce and Snecma assured that a dysfunctional thrust reverser on one engine would have by no means endangered the flight. At 16.39 finally, the Concorde was pushed back from the gate and began taxying towards runway 26R. The aircraft with a fully loaded passenger cabin had a total of 109 souls on board, the calculated takeoff weight was 185.1 tons. Next to Christian Marty, who had flown Concorde only 317 hours so far, sat first officer Jean Marcot (50). He had already been flying the supersonic jet since 1989, had raked up 2700 hours on it and also worked as a training captain. Behind them was flight engineer Gilles Jardinaud (58), also very experienced with over 12,500 flight hours, 937 of which on Concorde. For the heavily loaded flight to New York they had been filling the tanks almost to capacity with 95 tons or about 115,000 litres of kerosene, but that was not unusual. For the distance from gate to runway

Concorde flight AF4590 on July 25th, 2000

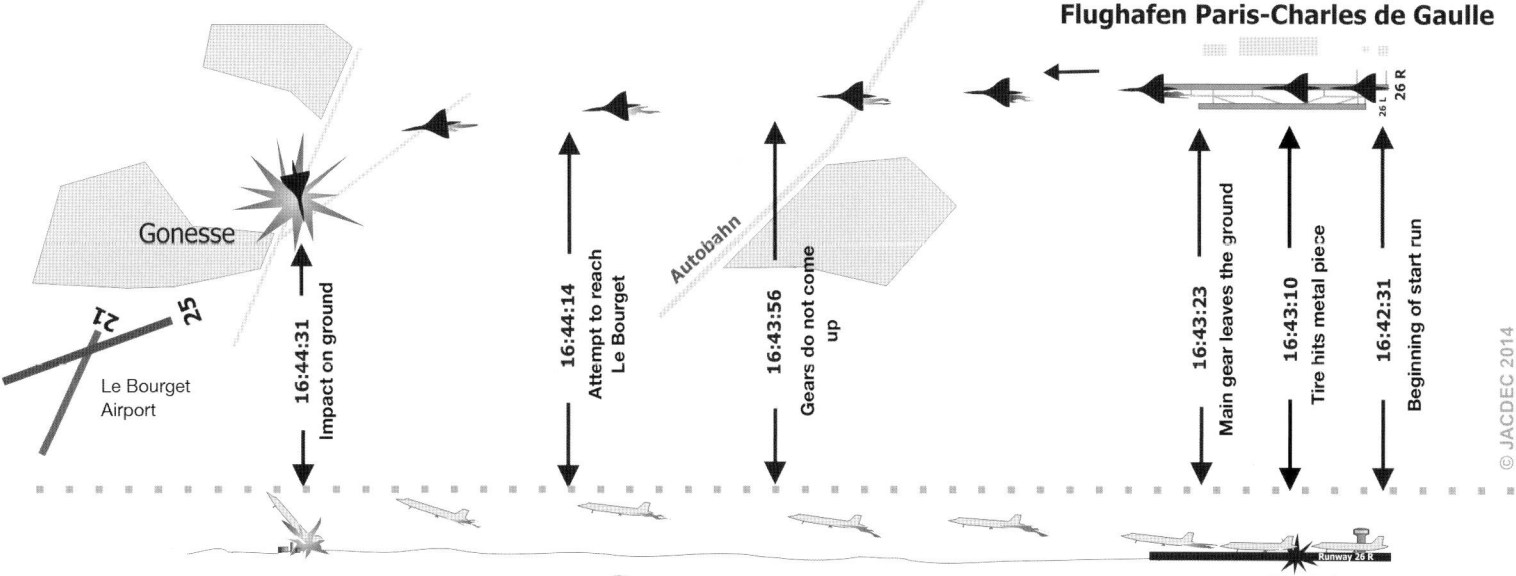

Flughafen Paris-Charles de Gaulle

The graphic depicts the sequence of events on AF4590 from right to left (JACDEC)

alone, Concorde used up about 800 kg or 1100 litres of fuel. No aircraft was more fuel guzzling. It was flying at a higher speed than a bullet, but Concorde consumed the same amount of fuel for a hundred passengers as a Boeing 747 with over 400 passengers flying in eight hours from Europe to New York.

Right in front of flight AF4590, a three-engine DC-10 from Continental Airlines had taken off from runway 26R, afterwards a Boeing 747 of Air France. Within sight of the Concorde pilots, another Air France Jumbo had just landed, a special flight carrying French President Jacques Chirac back

from a state visit to Japan. At 16.42 hours and 17 seconds local time or 14.42 GMT, the air traffic controller gave the takeoff clearance. In the following is an abbreviated and edited version of the recordings from the Cockpit Voice Recorder:

14 h 42 min 54.6 s, First Officer (FO): **"One hundred knots"**

14 h 42 min 57 s, Flight Engineer (FE): **"Four greens"** (meaning all engines run smoothly)

14 h 43 min 03.7 s, FO: **"V1".** (V1 is the speed, in this case 150 knots/270 km/h, when takeoff can't be aborted anymore and

the aircraft has to rotate)

14 h 43 min 13.0 s, FE announces: **"Watch out"** (a light vibration is sensed, at 14.43:10 the front right wheel of the left main gear has blown out)

14 h 43 min 13.4 s, Air Traffic Contol (ATC) tells the cockpit: **"Concorde zero … 4590, you have flames, you have flames behind you"** (the air traffic controller is so overwhelmed that he initially addresses the flight with number 001 used by scheduled services)

14 h 43 min 16.4 s, FE: **"Stop"**

14 h 43 min 20.4 s, FE: **"Failure eng... failure engine two"**

14 h 43 min 22.8 s, fire alarm sounds (engine number two has failed, while the warning system for burst tyres shows no irregularity. Unidentified voice over the radio saying "it severely burns")

14 h 43 min 24.8 s, FE: **"Shut down engine two"** (a procedure that is forbidden on takeoff below 400 feet /ca. 130 metres of altitude, especially as it hasn't been ordered by the Captain here)

14 h 43 min 25.8 s, Captain:

The ill-fated F-BTSC a few months before the crash at CDG in early 2000 (AS)

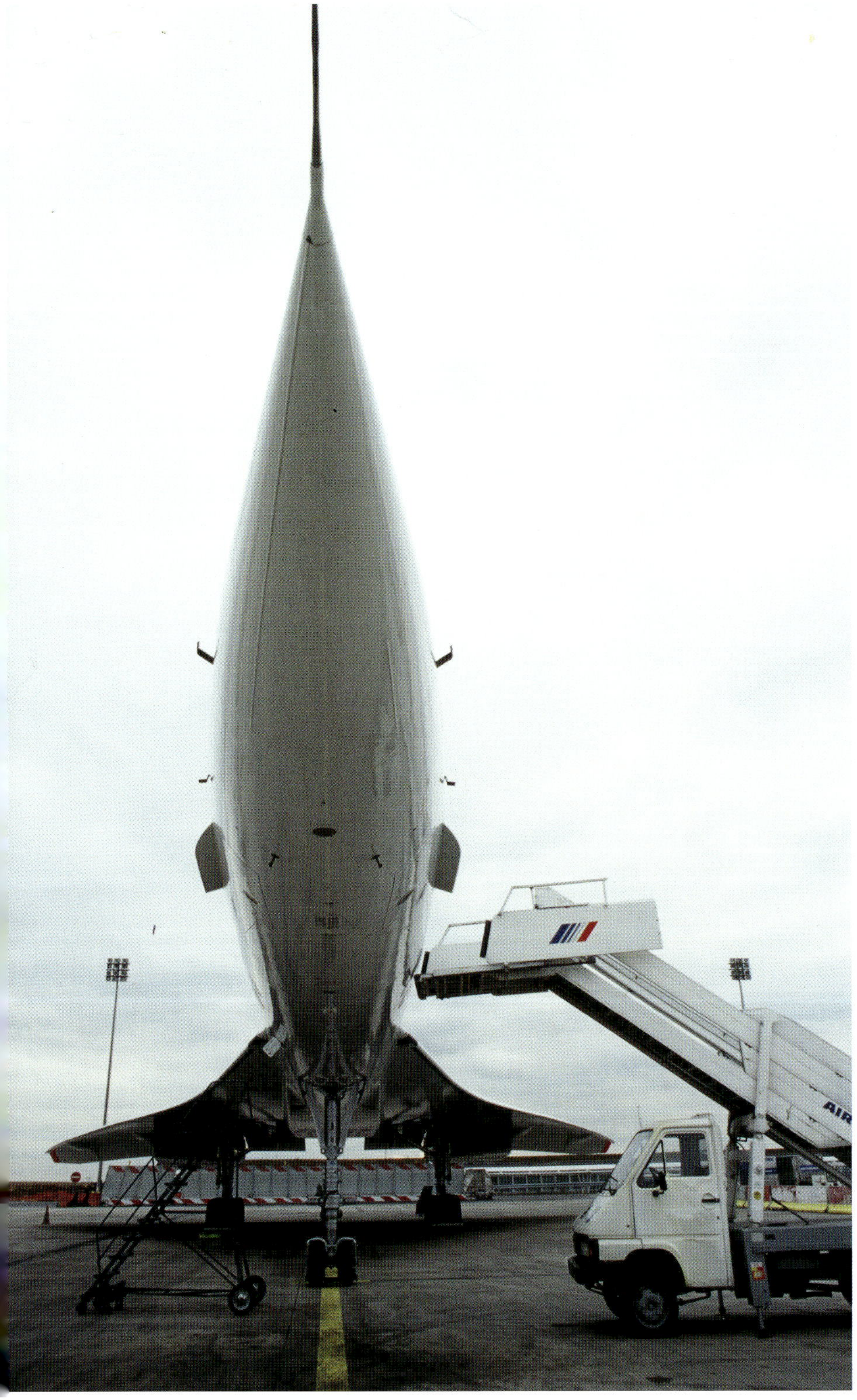

"Engine fire procedure" (*in the following second, noise of a selector, fire alarm stops*)

14 h 43 min 27.2 s, FO: **"Watch the airspeed, the airspeed, the airspeed"** (Concorde is flying at just 371 km/h at this point whilst 420 km/h would be necessary, therefore it passes the airport boundary at an altitude of just 30 metres)

14 h 43 min 29.3 s, fire handle pulled. New voice from the control tower: **"It's burning severely, I'm not sure if it's the engine."**

14 h 43 min 30 s, Captain: **"Gear on retract"**. (*In the course of the following eight seconds the crew mentions the landing gear several times. The Captain gave the order to retract the gear to increase speed*) ATC tells the cockpit: **"4590, you have strong flames behind you."**

14 h 43 min 42.3 s, second fire alarm (acoustic warning, the gear can't be retracted)

14 h 43 min 45.6 s, FO: "(**I'm trying**)", FE: **"I'm firing it"** (*meaning the fire extinguisher in engine number two*)

14 h 43 min 46.3 s, Captain: **"(Are you) shutting down engine two there?"** (*end of the smoke alarm, the engine seems to run again intermittently after a break*)

14 h 43 min 48.2 s, FE: **"I've shut it down"**

14 h 43 min 49.9 s, FO: "The airspeed" (Sound of a selector, end of acoustic warnings. The Concorde is still slower than 400 km/h and flying below 45 metres)

14 h 43 min 56.7 s, FO: **"The gear isn't retracting"**

14 h 43 min 58.6 s, third fire

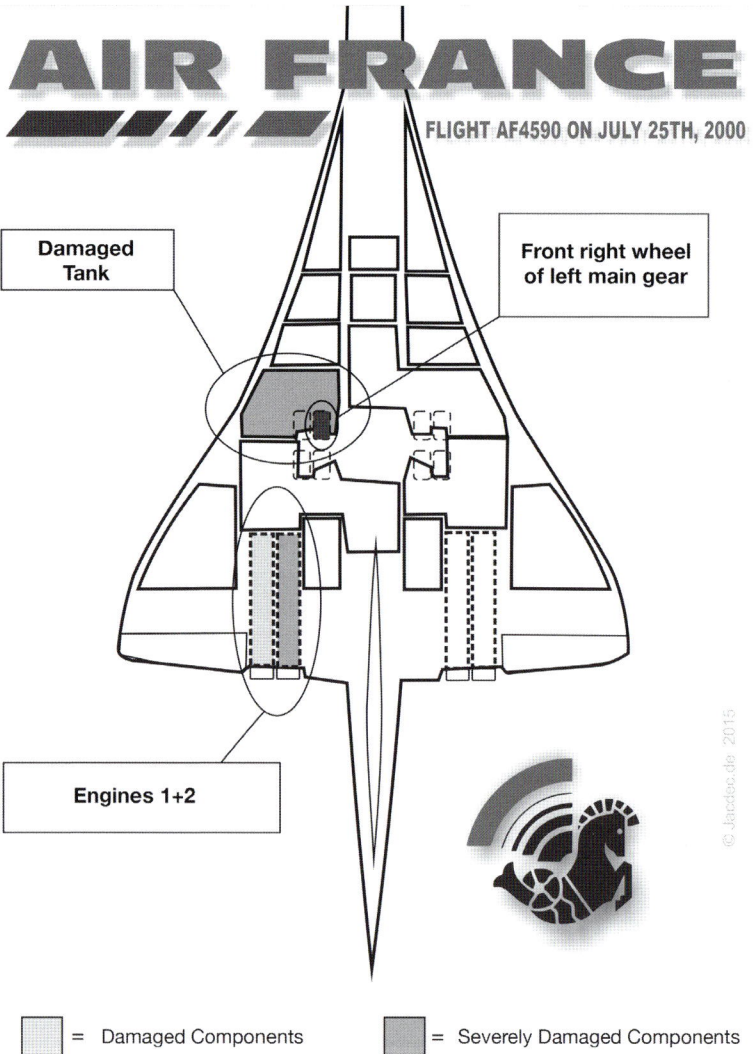

AIR FRANCE
FLIGHT AF4590 ON JULY 25TH, 2000
Damaged Tank
Front right wheel of left main gear
Engines 1+2
© Jacdec.de 2015
= Damaged Components
= Severely Damaged Components

The graphic shows the location of the damaged tanks on board the accident aircraft (JACDEC)

alarm. (*Besides the fire warning, the Ground Proximity Warning System also issues an acoustic warning. The aircraft has been in the air for 89 seconds*)

14 h 44 min 03 s, FO: **"The airspeed"**
14 h 44 min 14.6 s, FO: **"Le Bourget, Le Bourget!"**. Then a few seconds later: **"Negative, we're trying Le Bourget"** (*This was in reaction to the instructions given to the fire chief by the controller. Paris-Le Bourget airport is only four kilometres away from Charles de Gaulle, for an emergency landing there only a 90° turn is required, while doing a 180° turn to go back to CDG seems too risky to the first officer*).

The last sounds were recorded by the cockpit voice recorder at 14 h 44 min 31.6 seconds, a second later the recording ends. Two minutes and 13 seconds after the takeoff run of Air France flight AF4590 on runway 26R at Paris CDG airport, the aircraft crashed into the low- cost hotel "Hotelissimo" in Gonesse, only 6.4 km from the airport. All 109 souls on board immediately died, as did four members of the hotel staff on the ground.

The Concorde catastrophe was no ordinary aircraft accident, a crash of the most elegant, expensive and fast airliner humanity has ever come up with had been deemed inconceivable before. It caused a shock wave all over the world that some have likened to the sinking of the *Titanic*. An icon, a symbol for the technical feasibility of human dreams had lost its innocence after almost a quarter century in the skies. Hours after the crash a blurry photo was going viral that had been taken by two Hungarian aircraft spotters at CDG airport: It shows Concorde, hardly able to get into the air, dragging behind it a huge fire trail up to 60 metres long. Shortly afterwards, more disturbing visual documents of the catastrophe were surfacing: The wife of a trucker had recorded the burning Concorde by pure coincidence while trying out a new video camera from the driving truck. Only two days after the accident the most dramatic photo of the catastrophe made it onto title pages around the globe: it shows the burning Concorde from the front, shortly after takeoff. A Japanese businessman, who had been waiting to depart on another aircraft, took the shot through a cabin window with a disposable camera, a popular device before the era of cell phone cameras.

Soon it became apparent how the fatal chain of events in Paris had unfolded. The decisive role was played by a 43cm-long riveted metal strip made of titanium. The DC-10 of Continental Airlines that took off shortly before Concorde had lost the cover of an engine thrust reverser on the runway.

Concorde had overrun it with the right front wheel of the left main gear, which led to an explosive tyre burst. It was unusual that a chunk of tyre tread, 1.5 metres long and weighing 4.5 kg, came lose and was sent at high speed into the underside of the wing where the full tanks were located. The tyre burst was the main cause of the accident, and it was not the first time. Between 1979 and 1983, the French air accident authority BEA revealed, tyre debris had punctuated Concorde tanks on six different occasions. In total there had been 57 tyre-related incidents, the risk of such a problem was 66 times as high on a Concorde than on an Airbus A340.

The impact of the massive rubber chunk on the outside of the wing had not punctured the tank during takeoff of AF4590, but its immense kinetic energy triggered a never-experienced shock wave in the kerosene that lead to it tearing a DIN-sheet size hole into the tank from the inside out. Big amounts of kerosene gushing out ignited, the source was never identified, possibly severed electrical cables. Due to the heavy fire that ensued, not only did engine number 2 fail, but there was also a 23-second power loss in engine number one during the critical rotation phase. The investigation also looked at why Concorde Sierra Charlie was veering strongly left during its takeoff run. It was a fact that in the left main gear, due to a maintenance error at Air France, a spacer between the wheels was missing, impacting directional stability. Speculations that this could have been a decisive factor in the accident were vehemently dismissed by all active Concorde pilots and authorities. But it still remained open to discussion whether this shortfall did force the pilots to rotate their massively handicapped aircraft even before the pre-set rotation

The head of maintenance at British Airways shows the new radial tire of Concorde in 2001, designed to prevent bursting (AS)

speed was reached. Otherwise they might have feared a runway excursion, thereby putting in danger the waiting Jumbo Jet with the French President on board. It is a fact that Christian Marty rotated the aircraft at only 188 knots (338 km/h), eleven knots below the mandatory minimum rotation speed. Once in the air, the aircraft was a lost case, due to the strong impact of fire and the failing engines so close to the ground, the manoeuvrability was very limited. At last, Concorde reared to a nearly vertical position and reached a maximum of 200 metres of altitude before stalling and crashing tail first over the left side.

But initially the accident was not the end of the Anglo-French miracle bird, operators Air France and BA as well as the European aerospace giant EADS (today Airbus), legal successor to Concorde's manufacturers, did not throw in the towel that quickly. The fuel fire was seen as the main reason for the fatal ending of flight AF4590, and that is where modifications were planned to suppress gushes of fuel streaming out of leaking tanks ever again in the future. The undersides of the tanks were fitted out with a lineage made from Kevlar and rubber, praised for a self-sealing function in case of a leak. Also new burst-resistant radial tyres were specially developed for the Concorde. But even investments

of high double-digit million Euro sums into the remaining Concorde fleet were not finally able to halt the end of its history. Just when the modified supersonic airliners were about ready to return to service, the terror attacks on the USA hit on September 11th, 2001. After the terror attacks, demand for tickets immediately collapsed, full fare round trips had cost up to today's equivalent of over £10,000 / US$14,000 €12,000 prior to 9/11. The hopeful new beginning of flights to New York in November 2001 could not distract from this fact. "The lasting effects of 9/11 and persistent economic decline were followed by war in the Middle East and the disastrous occurrence of Severe Acute Respiratory Syndrome, or 'SARS.' The consequent sharp decline in demand for premium class travel had a particularly heavy effect on Concorde," recalled former BA CEO Baron Marshall in 2009.

For the time being, both airlines only served a limited flight schedule: BA departed daily except Saturday (later daily) at 10.30am from Heathrow to JFK, where the arrival was set at 9.25am local time. The return flight of the same aircraft took off at 12.15pm and was supposed to reach London at 9.10pm, too late for onward flights to the European continent. As soon as modifications of the fourth and fifth aircraft had been finished,

there would have supposedly been two daily flights again to and from JFK, among them a departure from JFK at 8.30am, as was announced in late 2001. "Five to seven per cent of our customers go to New York on Concorde, hold meetings for a few hours and fly back to London on Concorde at lunch time," reported Michael Bannister. A typical Concorde customer was

Concorde tanks in 2001 are now protected by Kelvar liners on their underside, designed to prevent puncturing (AS)

the CEO of a Munich-based advertising agency who had flown Concorde for 20 years and had done 30 to 40 flights on it annually in the years before the accident alone. She mostly flew in the morning from Munich to New York via Paris, arriving at a photo studio in Manhattan at about 10am local time, before taking a Jumbo Jet home that same evening. "Also the BA evening Concorde from London to New York was ideal, you could still do a dinner appointment in New York, spend the night and work the whole next day," the businesswoman told the author in an interview in 2001. "It is my personal feeling that work in specific professions gets spurred by the speed of Concorde." Especially in the fashion industry, supersonic tickets were part of the travel expenses, for herself, but also for the models. "They mostly flew on Concorde because their clients had to pay for their travel time. With Concorde, they had to pay less time and still saved money," she explained the rationale. Business started again after the hiatus. Between December 2001 and April 2002, BA again offered its popular Saturday rotation to Barbados in the Caribbean, scheduled flight time three hours and 45 minutes. But with a global economic slowdown, which mostly impacted bookings of expensive tickets and lead to massive losses at BA, the planned second rotation between London and New York was put on hold for the time being. Only by May 2003 should there have been a decision about expanding the supersonic offerings again.

The French initially only flew five times weekly (every day except Tuesday and Saturday) to and from New York, and they only sold a maximum 92 of the 100 seats on board, not until July 1st, 2002 were there were daily services again. The booking situation evolved promisingly at least for BA in the first month: between mid October and mid November 2001, almost 8000 Concorde tickets were sold, valued at almost £34 / US$48 / €40 million in today's value. That was almost the same amount as BA had supposedly earned in annual profits with Concorde before the accident, according to British media reports. Fares significantly increased for supersonic flight. Round trips from/to Germany for example with Concorde from Paris or London were now available from 5239 on Air France and cost from 5554 at BA, today's equivalent starting at about £6000 / US$8300 / €7000. Departures originating straight out of one of the Concorde hubs often were even more expensive. While load factors were initially increasing as well, chances for a longer term survival of supersonic operations did not seem that bad.

But Concorde also produced ever more negative headlines: In April and July 2002 there were two cases of engine surges, followed by shutdowns of the turbine affected, during supersonic cruise after which the handicapped aircraft had to descend from their cruise altitude of 50,000 feet (about 16,600 metres) to 30,000 feet (about 10,000 metres) and limp to their destination with subsonic speed – or even turn back. There was no danger for either aircraft or their passengers. "It's like a coughing of the engine, it's happening in a split second and rattles the aircraft quite a bit," explained BA Concorde chief engineer Claud Freeman. After the July 2002 incident the engine affected was disassembled on the ground and it showed damage. "If that has been triggered as a reaction to the surge or after foreign object damage on takeoff is still unclear," Freeman said shortly after. Also two aborted takeoffs brought Concorde bad press, one of which occurred in March 2002 with a cabin full of celebrities heading to Liza Minnelli's wedding. "For Concorde, public attention is always a lot higher than for any normal aircraft," acknowledged Claude Freeman. Jean-Louis Chatelain, former Air France Concorde pilot, explained: "After such an event, any small daily malfunction that is likely to normally occur in the aviation business would bring about much nervousness." And the incidents became more frequent and more serious. An old problem came back to haunt the aging diva, delamination at the bonded honeycomb structure of the tail's upper rudder. This led the whole bonded aluminium skin to peel off during flight, the whole section of the upper rudder tore off repeatedly while in the

Test campaign for the new radial tires on a French military air base (Michelin)

Big reception in New York JFK: In November 2001 Concorde's return is celebrated after the accident and 9/11, for the first time since 1977 AF and BA aircraft are put nose to nose (AS)

air. That had already happened several times in 1989, all aircraft had their rudders replaced as a precautionary measure. But in late 2002 the problem hit again, this time a BA Concorde lost the lower rudder section, while in February and May 2003, the same happened again twice on Air France flights. Not actually dangerous, but surely a warning sign. In February 2003, Air

France Concorde F-BTSD even had to divert to Halifax on the way from Paris to New York after an engine had to be shut off after a fuel leak was detected. The effect was immediate. "I think this was the moment when people at Air France thought: We've had enough. We don't want a second accident," said former BA Concorde pilot Christopher Orlebar.

Then there was another case with lots of adverse publicity, when British chancellor of the exchequer, later to be Prime Minister, Gordon Brown was on his way to a G7 summit in the USA on board a BA Concorde. Above the Atlantic, an engine began surging, the flight had to slow down and arrived with a delay of 30 minutes. Such surges during supersonic flights

did not happen often and were a concern. "To endure another major incident, or even a series of minor ones, threatened Airbus's reputation, it was another nail in Concorde's coffin," Orlebar pointed out.

Airbus indicated it wasn't willing to risk its image by an inherited aircraft of such extremes as Concorde. As for all of the aircraft major inspections

were due soon, Airbus declared the cost for these checks would be prohibitively high, indicating it might be as much as £40 million (about £65 / US$90 / €75 million in today's worth) for BA alone. "Airbus, with its reputation on the line and a new aircraft imminent (the A380) was not prepared to maintain support of Concorde at the old price," wrote Orlebar. They wanted a

way our modern life is – we are ruled by stakeholders. The operations were complicated and the economic situation was unfavourable," asserted the former Air France Concorde Captain. "The crash was of course an element that was important in the final decision. With such a small fleet, you cannot afford a significant adverse event, not to mention a crash."

On April 10th, 2003, BA and Air France simultaneously issued the announcement that they would end the supersonic era at the end of October 2003. "Concorde has served us well and we are extremely proud to have flown this marvellous and unique aircraft for the past 27 years," said BA CEO Rod Eddington. "This is the end of a fantastic era in world aviation but bringing forward Concorde's retirement is a prudent business decision at a time when we are having to make difficult decisions right across the airline. The decision to retire Concorde has been based on a long-term revenue and cost trend rather than recent events such as war in Iraq," stressed the airline boss in London. Meanwhile his counterpart, Air France CEO Jean-Cyril Spinetta was appearing in Paris declaring: "Air France deeply regrets having to make the decision to stop its Concorde

clear-cut exit for Concorde, the former pilot reckoned, and they had decided to take Concorde out of service earlier rather than later, "it was political." And he asserted that pressure had come from Air France and Airbus was a willing ally, deliberately setting the cost for maintaining supersonic flights at a high level. His former pilot colleague Jock Lowe even believed that

the Brits would have even been able to continue flying Concorde single-handedly. "Air France had lacked enthusiasm for Concorde since some time and now they found people at BA for the first time with open ears," said Lowe. Air France itself did not see fit to bear higher costs in times of tight business and under pressure to lower costs before the upcoming

privatisation. The British were not prepared to soldier on alone either. That's when Airbus pulled the plug and told both operators that its technical support for Concorde would end in October 2003. "It was an aging aircraft and we knew that," said Jean-Louis Chatelain. "It had to stop one day. The costs were rising, maybe sometimes unnecessarily. But this is the

Farewell flight of F-BVFB from Paris to Karlsruhe/Baden on June 24th, 2003, before the aircraft was taken to Technik Museum Sinsheim (AS)

operations but it has become a necessity." And he went on: "Maintenance costs have substantially increased since its return to service. Operating Concorde has become a severely and structurally loss-making operation." The much bigger verbal embrace of Concorde from the UK than from France was striking, with the statement of Airbus CEO Noël Forgeard being another point in case: "Airbus'

predecessors Aérospatiale and British Aircraft Corporation created Concorde some 40 years ago and we are proud of this remarkable achievement. But its maintenance regime is increasing fast with age. Thus, as an aircraft manufacturer, we understand completely and respect the decision of British Airways, especially in the present economic climate." This statement actually turned the true reasons kind of upside

down, as it had been Airbus putting a gun at the airlines' heads leaving no doubt they wanted to end the troublesome inherited Concorde adventure as soon as possible. Air France happily went along and only delivered a surprise in the last line of its press release with the sad announcement that supersonic flights were already ending in seven weeks' time on May 31st, 2003.

In the UK, they did not want to

give in to the end of supersonic flying quite as hastily. BA intended to use up all the time until the deadline Airbus imposed to the very last day. "Our pride in the aircraft will never wane and I am determined that we make its final six months in the sky a time for celebration," Eddington said, resisting the assiduous obedience the French were happy to demonstrate. And at the same time he took the opportunity to announce

special Concorde fares for the coming months: £1999 for a transatlantic trip, one way on Concorde and return in Economy, and £3999 for a Concorde return. This equals about £3230 / £6460 today, or up to US$9000 / €7500. Air France CEO Jean-Cyril Spinetta was forced to offer a justification when asked about the abrupt ending of supersonic flights: "Links between London and New York have always generated more substantial traffic flows than between Paris and New York. British Airways has therefore more reason to operate its Concorde flights longer than Air France." The announcement of an imminent end to supersonic flying created a never-before seen demand for the remaining Concorde flights, whose tickets had suddenly turned into goods of limited supply. While supersonic flying had so far been an elitist and somehow subdued affair, there now became a jubilant atmosphere during the last months. Even many people with limited means scraped together their savings to do a last and only, unforgettable Concorde flight, mostly with a companion. So, no wonder the atmosphere was much noisier and joyful than before, and champagne was flowing even more freely than before.

On May 31st, 2003, the last scheduled Air France flight of the supersonic era landed at

October 24th, 2003 marks the last day of scheduled Concorde flights of British Airways (BA)

Air France Concorde F-BTSD landing for the last time in Paris Le Bourget on June 14th, 2003, where it remains on museum display ever since (AS)

The final roar – the end of Concorde is featured on the covers of most industry publications (AS)

On June 24th, 2003 Air France Concorde Fox Bravo has its last landing witnessed by 20,000 spectators at Karlsruhe/Baden (Technik Museum)

5.44pm local time at Paris CDG airport from New York JFK, carrying just 68 passengers. Thousands of spectators were crowding the taxiway fences, some of them waving the Tricolore. When the passengers disembarked, hundreds of Air France employees were lined up to bid them and Concorde goodbye. There were no lavish farewell ceremonies, as this did not seem appropriate to management even three years after the accident. A few days later there was a special flight from Paris to New York and back with invited guests, the most high-level frequent fliers and government representatives. After that, all four Air France Concordes flew a last time to start off their next life at a museum. The flights went to destinations near and far, from Paris Le Bourget and Toulouse to Washington DC. On June 24th, 2003, F-BVFB took off from Paris and initially overflew the Bay of Biscay in a supersonic loop, then on to Karlsruhe-Baden airport in southwest Germany, where it touched down in front of thousands of spectators for its last, 5473rd landing. In the following weeks the aircraft was partially dismantled by experts and transported by road and river barge to Technik Museum Sinsheim near Heidelberg, where it was re-assembled and put up on a museum rooftop. Mounted there in its spectacularly steep takeoff angle, in synchronicity with the Tupolev Tu-144 registered CCCP-77112 standing next to it in similar fashion. Creating a unique mecca for the first supersonic era, as this is the only place in the world where one can inspect both designs from close quarters. After its last landing, Guy Tardieu, deputy CEO of Air France, remembered the aircraft's darkest days: "The tragedy of July 25th, 2000, will always remain in our memories. The homage that we owe to the victims of this drama that touched our German friends so profoundly brought Air France to the decision to preserve its Concorde fleet and let it become part of the world cultural heritage after the decision to end its operations. We deemed it natural therefore to make available one aircraft to the Technik Museum Sinsheim. It is a great pleasure for us to deliver Concorde Fox Bravo to the museum, which herewith gets a second existence and can be admired by the greatest possible number of visitors." On June 27th, 2003, the very last landing of an Air France Concorde ever took place at its birthplace in Toulouse, where Fox Charlie is on display today in front of the Aeroscopia museum.

Meanwhile BA maintained its scheduled services to New York as well as five summer rotations to Barbados, the last one on August 30th, 2003. The BA fleet saw particularly heavy utilisation in October shortly before the final decommissioning. Initially its Concordes were sent on a mini tour through North America with flights to Toronto, Boston and Washington DC, followed by a UK farewell tour during the very last days with landings in Birmingham, Belfast, Manchester and Cardiff. And then the grand finale on October 24th, 2003, at London Heathrow, when shortly after 4pm all of the last passenger flights came in for landing. One Concorde from Edinburgh, one after a flight over the Bay of Biscay and finally, at 4.05pm, the last scheduled flight BA002 from New York. "The eagles have landed. Welcome home", that's how the air traffic controller in the tower ended the era of Concorde passenger flights after the last touchdown. At the same time, Concorde chief pilot Michael Bannister told his last passengers on board: "Concorde was a fabulous legend born from dreams, built from vision, operated with pride by a family that loved her. A fantastic aircraft that became a legend today. Thanks for flying British Airways Concorde." After a grand reception for the disembarking, sometimes prominent passengers, BA arranged for all five active Concordes to be assembled on the apron for a farewell photo shoot. Followed by a big party for the "Concorde family". *The Independent* on the following morning chose the headline: "At 4.07pm, one of the world's most exotic birds became extinct."

In the following weeks the delivery flights for six of the supersonic jets took place to their final resting places in museums from Manchester to Seattle. On its journey to the Museum of Flight in Boeing's hometown, Concorde Alpha Golf set a final record: with special permission it was allowed to fly supersonic over land from its intermediate stop at New York JFK over sparsely populated areas in Canada to Seattle, and covered the distance in record time of three hours, 55 minutes and twelve seconds. Shortly after, the last Concorde flight of all times was due. On November

26th, 2003, almost 35 years after the first flight in Toulouse, Alpha Foxtrot touched down shortly after 1pm at Filton, the Airbus company airfield in Bristol. The flight was fully occupied by a hundred airline staff, about 20,000 spectators were gathered around the airport. Prince Andrew officially greeted the last Concorde flight ever after landing: "Today is one of the saddest days in aviation history, but at the same time, it's a day to reflect on the glory of what the UK can achieve."

After all the farewell emotions, one problem remained unsolved, the legal clarification of who was at fault in the Concorde crash; the proceedings dragged on for long even after Concorde flights had ended for good. Air France had already settled out of court with the victim's families in 2001, agreeing to pay US$150 million in compensations and damages. But the investigators of the French judicial system pressed on and amassed 80,000 pages of documents. Employees of three institutions were in focus of the investigations: Air France, Continental Airlines and the French civil aviation authority DGAC. In July 2008, a prosecutor in the Paris suburb Pontoise, decided that criminal charges would be brought against Continental Airlines as the only airline, the company and two of its mechanics who allegedly were responsible for the faulty maintenance work

on the engine part found on the runway. In addition, the same charges were brought against two formerly high-ranking engineers of Aérospatiale, against the first director of the Concorde programme and the former Concorde chief engineer. The former director of technical services for Concorde at the DGAC was accused of negligence, as he allegedly had ignored warning signs such as a series of tyre bursts over a 15-year period. The long and arduous trial proceedings took from May to December 2010 before a ruling was handed down: none of the French defendants was punished. Aérospatiale or rather its successor, Airbus, was ordered to pay a portion of the damages.

Continental Airlines, and also the mechanic involved, were found guilty of involuntary manslaughter, and while his supervisor was acquitted, the mechanic was sentenced by the court to a 15-month suspended sentence. Continental Airlines was held liable for damages of €1 million to Air France for damaging its reputation. As to be expected, the US attorneys protested against the ruling blaming solely the

American defendants, they got unexpected support from the lawyer of the deceased Air France pilot's family: "I don't understand the difference of treatment between the American transporter and those who were involved on the French side," he stated. Appeals were filed and in November 2012, an appeals court judge overturned the guilty verdicts and acquitted both the airline and its mechanic. Even if they may have made mistakes, they would not be criminally responsible for the deaths. But she upheld the finding that the dropped metal strip had been the precipitating cause of the accident and therefore confirmed Continental had to pay damages to Air France. At the same time she directed harsh criticism at the French authority DGAC for doing nothing in "25 years of operation … littered with numerous cases of tyre damage following more or less serious incidents." As these were happening repeatedly, the judge said the authorities should have suspended Concorde's airworthiness certificate. This, however, only happened weeks after the Paris crash. In December 2013, the accused Continental Airlines mechanic as "the last victim of the Concorde crash," as his attorney referred to him, tried to get absolution in filing a suit against Air France-KLM and United Airlines that had incorporated Continental. The

G-BOAC on its last flight from Heathrow to Manchester, October 31st, 2003 (BA)

lawsuit was not able to proceed and the life of the mechanic was devastated in any case, even if he and Continental were eventually cleared of all charges of involuntary manslaughter. So in the end, no one was really held responsible for the death of 113 people on board and on the ground.

But over time, the myth of the timelessly beautiful Concorde as a remarkable achievement of technological progress and human ingenuity gained the upper hand in public perception again. The dark shadow hanging over it since the accident faded away slowly. Especially as the course of events on that fateful July day had nothing to do with the fact that it involved a supersonic aircraft, but rather with an unfortunate chain of events depending on each other, as in most other aircraft accidents. Even as it had become exposed that both oversight authorities and airline operators had not made sure that known safety deficiencies were addressed in time, but rather only patched over by short-term actions.

And that they had always shied away from bigger investments that possibly could have prevented the accident in July 2000.

In the end, positive views of Concorde prevailed. "All of us involved in the programme were helped in our task by three rarely united factors," commented former Sud Aviation chief test pilot André Turcat (1921-2016) in 2003 who had flown Concorde on her first flight. "Our bosses were professionals and understood our problems and technical exigencies, our governments helped us without hesitation and we had the support and enthusiasm of everyone – not least the general public." British politician Tony Benn (1925-2014) had always supported Concorde, also while being minister and member of Labour governments. "Concorde led the world and still does since neither America nor Russia with all their skills and resources was ever able to build a supersonic airliner and operate it as Air France and British Airways have done so efficiently,"

Benn summed it up in 2003: "I recognise – very reluctantly – that it cannot go on forever and the last flight will be very emotional for me as it will be for many thousands of scientists, engineers and craftsmen who built it." On the event of the 40th anniversary of the first flight in 2009, veteran industry observer Pierre Sparaco summed it up very fittingly: "Concorde's legacy nevertheless stands as a monumental achievement. Very few regret the money and man-hours spent."

Five Concorde at once at the British Airways Heathrow base after the last flight on October 24th, 2003 (BA)

Farewell photo of British Airways Concorde staff and its five aircraft on October 24th, 2003 (BA)

eleven

"Concorde never had a market"

Talking Supersonic Travel with former Air France-Concorde captain Jean-Louis Chatelain.

Jean-Louis Chatelain in 2020, posing in front of his 2003 portrait at Technik Museum Sinsheim (AS)

Jean-Louis Chatelain, born 1946, flew Concorde as Captain on the left seat between 2001 and 2003. "I was certified to fly Concorde aged 54, receiving my type rating exactly the day before the accident in July 2000. The crash was a triple shock for me", explains Chatelain, as he also lost close friends in the perished crew.

The pilot later spent a whole year working in the French investigation commission to find the causes of the Concorde accident, while his actual flying career on the aircraft was cut short by the events. "Normally I would have flown the aircraft for four years, but I had only 18 months with it in the end." He flew a total of 400 hours on Concorde, 260 of which were in supersonic flight; his flight log records 92 transatlantic legs with Mach 2. He flew many supersonic charters above the Bay of Biscay and a check flight after major maintenance with a deliberate shutdown of one engine at Mach 2 above the Atlantic.

Jean-Louis was at the controls for the memorable last flight of Air France F-BVFB "Fox Bravo" on June 24th, 2003, flying from Paris to Karlsruhe/ Baden in Germany, for the delivery to its last resting place at the Technikmuseum Sinsheim near Heidelberg. It has been on display ever since on the museum roof alongside its Soviet rival Tupolev Tu-144, the only place in the world where both aircraft can be seen.

This interview with Jean-Louis Chatelain took place on location at Sinsheim.

What were the biggest achievements of Concorde?

The achievements of Concorde in hindsight were better than expected. It was a total challenge at the beginning, but they achieved their technical objectives and it worked perfectly for 25 years with advanced technology like fly-by-wire and carbon brakes.

It was the first cross-border project ever. It was an international adventure with the French and the British, and we learnt a lot from that. It was the start of Airbus and wider international projects in general. For everybody, having this Concorde experience was really an advantage. Concorde taught the French how to run and manage an international project. Airbus would not be Airbus without Concorde. It was such an advanced aircraft, Airbus just had to make an extrapolation of the insights gained and progress made with Concorde. Of course, fly by wire for Airbus was not based on analogue computers, but on digital ones.

How does Concorde compare with other aircraft you flew in your career?

The Boeing 707 was really a raw aircraft, but getting back from an advanced technological aircraft such as the A340 I flew before to Concorde was not a problem, the basics were there. What I had to learn was supersonic flight, as I had no military background. You have to adjust your flight planning to the speed. You move from 8nm per minute to 20nm per minute.

Was there ever a realistic chance for supersonic travel becoming a major market?

Concorde was not based on a business plan. The decision was taken at French government level to run as an advanced project – as was the decision for nuclear submarines and the establishment of the space centre in French Guyana. Speed has a cost on trip fuel figures. So, if you take the marketing approach, it was due to fail. And it was intended to fly supersonic over land, which was not allowed then because of the sonic boom. In the end, the Americans took the logical decision to pull out, and the British and French kept this big project. But if you look at the figures, they show the market was not there, it was never there. During my time we only had about 30 to 40 passengers on every Concorde flight.

In the late 1990s there were plans to extend the life of Concorde until the second decade of the new millennium.

Why, and was that realistic?

I do confirm that it was the plan to extend the life of the aircraft from the airworthiness point of view, and before the crash there were already negotiations to extend the life of Concorde until 2017. You cannot measure the image benefit of flying Concorde, both for Air France and British Airways, but it was a real benefit. We had people flying Concorde that were frequent fliers in First Class because they flew Concorde from Europe to the US. It brought additional revenue in this way. And because of the image the aircraft was carrying, it was adding to the value of the brand of the two operators. From that point of view, it was not unthinkable to extend the life of the aircraft.

Can you sum up the sequence of events that led to the accident?

I was a member of the accident investigation team and it took one year of my life. The scenario was a sequence of events like in most aircraft accidents. The triggering event was not proportional to its consequences. Here it was a 200-gram piece of metal that made a clear cut into the tyre at high speed. That provoked a tyre burst, with a huge piece of tyre impacting the wing, and then you had this shock wave developing in the tank, and the tank ripping open from the inside to the outside. Causing a massive fuel leak and a blaze of fire, leading to the loss of power in two engines, which no aircraft is certified to withstand. We were actually able to objectively reconstruct the sequence of events. Witness accounts that there was a fire before the tank burst were honest, but wrong, we all know how human witnesses can err. The fact all this happened in that way was incredible, and very random. They tried to reproduce the cutting of the tyre in the investigation. They found that, depending on the angle of attack of the tyre on this little metal piece, it would cut or not. Once two engines had failed, it was impossible to recover the aircraft because of the characteristics of Concorde's delta wings. The drag is so tremendous at low speed that you need extra power to get above the critical speed. If you don't have that power coming from the afterburners, you are lost as you lose speed. That was the situation they were in.

Concorde suffered many incidents, before and after the accident. Was it a safe aircraft?

The answer is yes. It was never considered that a burst tyre could trigger such a shock wave inside the tank, this was learnt after the crash only. But it is true that the tyres were a weak point in this aircraft. The Washington incident was serious already, but nothing to be compared to the accident scenario.

Did the modifications after the accident improve Concorde's safety?

Definitely. The new radial tyres were amazing. We had two incidents after the introduction of the radial tyres, one at each airline. In both cases the tyres ran over metal parts, in the case of British Airways it was parts of an oxygen bottle for whatever reason. The aircraft was able to take off and land, with tremendous stress on the tyres, while the piece of metal was still embedded in the tyre. That tells a lot of the improvement the new tyres and its technology brought. For the new Kevlar lining of the tank interior, there is no proof of its effect, but I assume it was an improvement. All the highly publicised incidents after the crash were nothing extraordinary, just that it got so much more attention then.

What can you say about the end of Concorde?

The decision to end Concorde's active life was taken not by the aviation authorities or

even Airbus, which had inherited the role of manufacturer from Aérospatiale, but on government level in both France and the UK, at the ministries of transportation. And of course, Airbus had other priorities with its upcoming A380 at the time than taking the assets of Aérospatiale and investing in the airworthiness of Concorde, for sure. But the ones who had to push through the decision were the airworthiness authorities. First, they suspended the type certificate for some 15 months, then they stated they would not issue a follow-up airworthiness certificate. This killed operations for both the British and the French. My understanding was that the airworthiness follow-up would have to be done much more on the French than on the British side.

Why was the French side so eager to stop Concorde, while the British were eager to keep flying?

It was much more difficult to recover from a crash for the airline who suffered from it, Air France. That meant such damage for the airline's image that it took so much time to rebuild the image of safety for the aircraft. But for sure the French decision to stop to follow up maintaining its airworthiness condemned the British to follow. Concorde was an aging aircraft for sure,

Jean-Louis Chatelain, born 1946, began his pilot career at Air France on the Boeing 707 and Caravelle. After the end of Concorde he took a stint as Airbus test pilot (AS)

meaning the cost was rising to mitigate more frequent and recurrent system failures, which meant higher maintenance costs, because there was much more work to be done. In that respect, Concorde was not so different than other aging aircraft.

What do you feel coming to the museum in Sinsheim,

seeing the aircraft you flew?

It's quite emotional of course. It is almost unreal, walking underneath this big aircraft, knowing that it did fly. And because this is part of aviation history and has never been replaced by anything. It looks somehow unreal, and I think to myself: Did it really exist? Did I fly this one? Sometimes Concorde

appears when I dream, that's why it has something unreal for me as well. I flew for eight years after the end of Concorde, but it was never near the same experience.

Do you think it would be good to have a supersonic airliner again in the future?

I don't think so. We have every other opportunity to work on

A total of 18 months Jean-Louis Chatelain flew Concorde between 2001 and 2003, accumulating about 400 hours (AS)

a very short time could be an acceptable cause for coming up with supersonic travel. But nowadays, we just lack a justifiable cause.

advanced technology. I think we have passed a landmark in aviation, we have exited the pioneering age, where we could claim progress in safety, speed etc. every year. It has become an industry for the benefit of most people. I think there is no cause for taking people at Mach 2 again. Nowadays, what would justify crossing the Atlantic in three hours instead of eight hours at the cost of a huge impact on the planet?

At the same time, we have every kind of technological means to communicate or to perform surgery even from remote places. The big human enterprises are based on a just cause. And I don't think there is any just cause anymore for supersonic flights. Before, we were in a phase of achieving more progress every year, and speed was seen as progress at the time.

When Concorde started service in 1976, I was based as a young Boeing 707 pilot in Lima, Peru, and I did not have communication with my family for three months. Understandably, not only was speed seen as a progress at the time, but the achievement of taking a flight from one side of the Atlantic to the other in

The French veteran pilot has close ties with Technik Museum Sinsheim since delivering the museum's Concorde to nearby Karlsruhe in 2003 (AS)

twelve

The long way to a new generation of supersonic airliners

"It is a major disappointment to me that Concorde never had a successor. Undertaking a project like Concorde without planning a follow-on is unfortunately senseless."

Pierre Gautier, former French Concorde programme director, 2003

"We are at a point where we believe we can build a quiet supersonic aircraft."

Peter Coen, NASA project manager, 2016

Boeing's Sonic Cruiser design (AS)

Successful aircraft are never loners, they are always conceived to be part of a family plan. There are improved, stretched or shortened versions, each for specific market demands. With Concorde initially an enhanced so-called B version was part of the concept. The two airframes MSN 217 and 218 had been earmarked to become B models, with widened wing tips and new slats for improved handling at low speeds. But this never came into production, after MSN 216 had been assembled in 1978, the 20th aircraft built in total including all prototypes and pre-production models, it was the end of the line. In 1981, the last Tu-144 was rolled out, and since then no supersonic airliner has been produced, four decades have passed without innovation in supersonic transportation. Since aviation's pioneering days, speed had always been a driving force in the evolution of building aircraft. Successive technological leaps from the *Wright Flyer* to the prototype of the Boeing 707 took less than 50 years of continuous progress. But in recent decades this momentum has come to a standstill. The last speed records date from the mid-1970s: in 1974 Concorde reached Mach

2.23 (2405 km/h), while in 1976 the US reconnaissance aircraft Lockheed SR-71 Blackbird reached a world record for aircraft with air-breathing engines – 3367 km/h or Mach 3.36. As today's airlines want to save fuel they are even slowing their aircraft down, a transatlantic flight takes longer today than 40 years ago.

Only in the second decade of the new millennium did fresh momentum build up again, even if in a much more modest way than in the 1960s. But it has become more and more likely that passengers will be able to fly on commercial supersonic services again in the late 2020s. It will have to be seen if the fares will be more affordable for wider parts of the public than for Concorde flights, or if it will just be a luxurious transport option for the super rich able to afford a supersonic business jet. At least science, research and production methods will reach a state at the end of this decade when new materials and the most recent research results will enable much more affordable, but most of all more sustainable and more economical supersonic flights. Materials already available in the 1990s for a new beginning were titanium, heat-resistant aluminium and polymer composite materials. In principle, 30 years on, what the former NASA Director High Speed Research Louis J.

Williams said in 1991 is still true: "It is all but an easy challenge, at this point it has to be viewed as a very costly and high-risk venture."

The question as to whether there would be a "new Concorde" has kept the media in particular busy for almost 40 years. Countless graphics of sleek models soaring into the air in front of a dramatic evening sky have been published by manufacturers and presented as possible successors. But, in stark contrast to the flashy publicity material, what really happened was absolutely nothing. Besides wind-tunnel models, no design ever reached a stage anywhere remotely close to building a real aircraft, there were just supersonic tests with modified fighter jets. There was a steady stream of funds, however, mostly in the US; where it seemed America tried to get absolution from the trauma of its earlier SST programme cancelled in 1971 after insane amounts of money had been spent in vain. It is almost ironic that all realistic supersonic projects currently come out of the US, but none of them being built by any of the big manufacturers or with any government funding. The one exception is the NASA-Lockheed Martin research aircraft X-59 QueSST (more on that in this chapter). The road to this point has been long and rocky. There was plenty

of researching, testing and designing, mostly in 1990s America, but also in Europe and Japan. But again, it seemed that the important question of what realistic market there was for all these great concepts people worked on got out of

focus. Mostly it was all about keeping an armada of engineers and consultants busy, being nurtured by NASA's well stocked research funds, as Ronald Davies was suspecting. Back in 1986, NASA had already launched its research

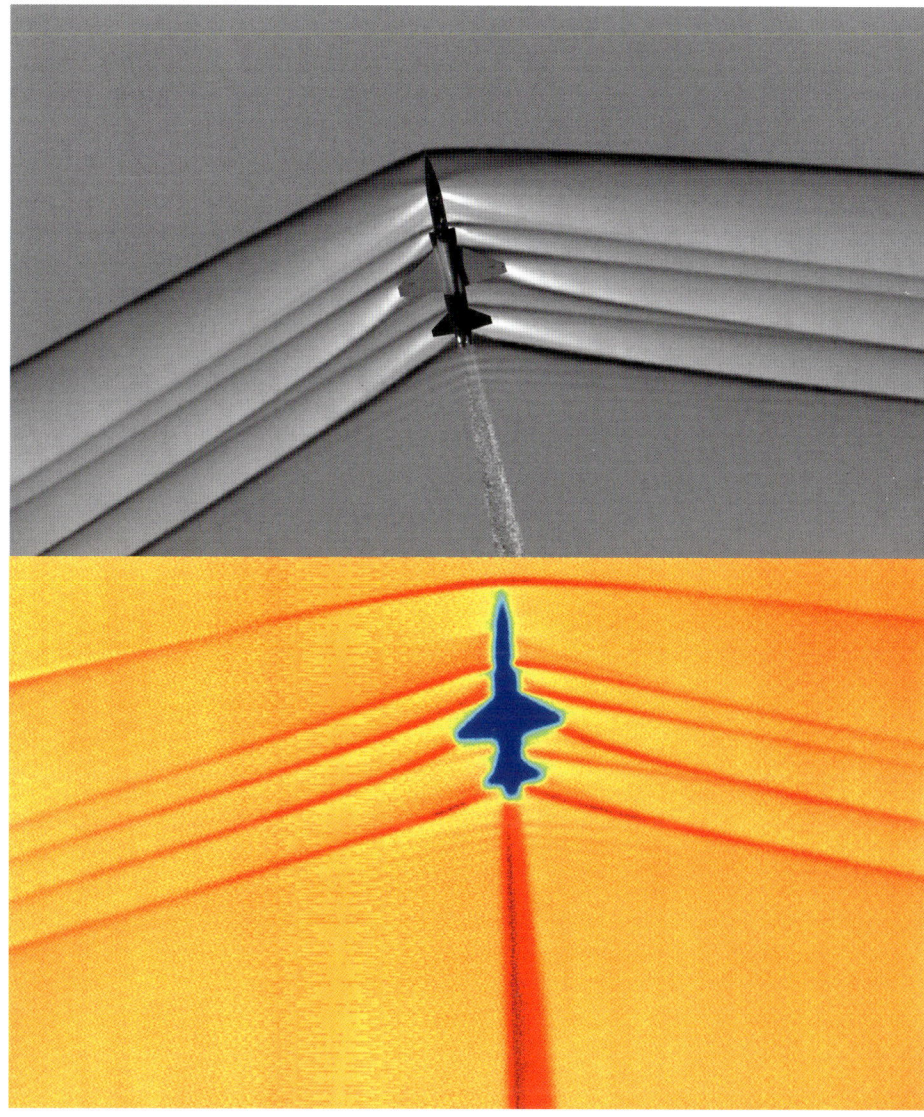

The so-called Schlieren method made it possible for scientists to generate a flow visualization of sonic waves generated by supersonic aircraft (NASA)

programme on a High Speed Civil Transport (HSCT), a politically correct euphemism, after the acronym SST with its negative connotation had been erased from industry language for the time being. From 1990, there were supposed to be two phases of research about sustainability and economic efficiency. All funded by a NASA budget of US$400 million to be spent over eight years.

A second supersonic aircraft generation would be possible, stressed NASA's Louis J. Williams in 1989. "But it probably won't fly until 2000, as the technology currently available is not sufficient and the public will turn against such a development if a new aircraft is not clearly much more sustainable." Supersonic flying with extreme noise around the airport, inevitable sonic boom

and substantial emissions in the higher, more sensitive altitudes such as Concorde flew at would not be tolerated by anyone anymore. In Germany, Lufthansa's head of engineering Rolf Stüssel had already clarified his objectives for such a project in 1990: "To operate economically and compete with today's long haul aircraft, it should have a range of 12,000-13,000 km and a capacity of 250-300 passengers". Stüssel stated: "The most important prerequisite that has to be ensured, besides a proof of economical priority, is environmental compatibility. In that area, there are two main problems: 1. Noise emissions on takeoff and landing have to be within the limits applied to conventional aircraft at the time. 2. The impact of the sonic boom has to be so limited that

the aircraft can fly supersonic also over land. 3. Engine technology has to ensure that emissions of nitrogen oxide are limited to a minimum." Stüssel laid out very clear airline demands.

Before the second phase began in 1993, NASA explainined its motivation, but also the risks. Hope was that it could be lucrative for the US, bringing a positive trade balance of US$200-300 million through sales, creating up to 14,000 highly skilled jobs and preserving the overall strength of the US aerospace industry. NASA estimated the necessary investments at US$4.5 billion even before the programme was launched. It suggested initially spending US$1.5 billion for the second phase of the research programme to examine technologies that were the

highest priority, and also those with the biggest uncertainties. "The objectives in any technology are very challenging and pose a great risk, but they are achievable," said NASA at the time. If the success needed in sustainability end economical feasibility was not achievable they would take a hard look at the programme, cancelling it altogether if necessary. One of the crucial points that had not been solved yet was described by Louis J. Williams in 1991: "We were able to conquer the speed of sound, but we haven't conquered the sound of speed yet." It is important to stress how vital it would be for the economical opportunities of such an undertaking to have an aircraft able to fly supersonic over land, which has been forbidden over the US since 1973. "This would increase the

potential market by 30 per cent or more," claimed Williams. But all market projections of that time, as well as the early era of the Concorde programme, have to be regarded with the utmost scepticism, as Ronald Davies did in his 1998 book *Supersonic (Airliner) Non-Sense*.

"Anxious not to look a gift horse in the mouth, Messrs Boeing and Douglas promptly complied, with vigour, and drew upon vast resources of wishful thinking and special pleading," analysed Davies about the HSCT activities, "each produced impressive-looking documents". The market outlook again, as decades before, lacked any reliable base. "Only a cursory inspection of the market estimates revealed considerable reliance upon vague assumption that bore little resemblance to the accepted parameters upon

In the early 2000s Airbus was conducting research on a possible Supersonic Commercial Transport (SCT) project (AS)

which all forecasts of airliner market numbers should be based," wrote Davies, who himself had earlier been a market researcher for McDonnell Douglas. The assumptions, he stated, "were as full of holes as the proverbial colander". In an elegant brochure Boeing announced around 1990 that it was seeing markets from North America to Asia (transpacific), North America to Europe (transatlantic) as well as Europe to Asia as suitable for an HSCT. That was questionable already due to the simple fact that for the latter route, supersonic flying over land would be unavoidable, but was not realistically to be expected. And transpacific is an equally questionable market due to some unavoidable realities. These have not only been pointed out by Ronald Davies (see Chapter 5), but also been brought to our attention by Lufthansa's head of engineering Rolf Stüssel in 1990: "As supersonic aircraft are mostly operated in an East-West direction or vice versa, for the choice of cruise speed, time differences around the globe play an important role". The Lufthansa engineer further outlined what would be needed for success: "High productivity for the airline and acceptable departure and arrival times with real time gains for passengers have to be harmonized". This is impossible to achieve with

Since 2002 Aerion Supersonic was investing in research for a small supersonic airliner (Aerion)

supersonic aircraft due to the existing time zones, as described earlier.

But that fact didn't prevent Boeing from drawing the conclusion: "Based on this projection, the demand in HSCT markets will be 315,000 passengers per day in the year 2000, growing to 607,000 in the year 2015. This is a potential market for 1,000 to 1,500 HSCTs – sufficient to justify further study". Boeing was looking forward to more millions of Dollars from NASA to keep its hundreds of ambitious engineers busy. Boeing's client NASA sang to exactly the same tune and did not hesitate to adopt such unrealistic projections without questions. The studies came to the conclusion "that one could realistically expect from a modern supersonic airliner to win a share of the international long-haul market that would warrant its development," seconded NASA's Louis J. Williams in 1991. He continued, "Conservative estimates of

Lockheed Martin's first draft for a quiet supersonic jet in 2012 (NASA)

the supersonic market for 2015 assume over 600,000 passengers a day". Davies left no doubt about his disdain: "Armed with a number of meaningless charts and empty statements, the astonishing conclusion was drawn that 'if 100% of the first – and business-class and only 25% of the economy-class passengers are captured, there is a potential requirement for over 600 Mach 2.4 vehicles.'" The chairman of Rolls-Royce, however, Sir Ralph Robbins,

dared to publicly declare in 1998 that the number of supersonic airliners needed would not actually exceed one hundred. "This number when set against the projected development cost does not make economic sense," it immediately dawned on Brian Trubshaw. Still, NASA continued to motivate itself, the managers responsible for structure and propulsion summed it up as follows in 1994: "Current small scale lab tests are very promising

in fulfilling economical and environmental requirements. With these technologies in hand the American civil aircraft manufacturers will be able to give the international public what it demands." Probably much closer to reality was the last sentence in the otherwise overly optimistic Boeing brochure about the supersonic airliner: "Boeing will build an HSCT when it becomes commercially viable." This of course never happened,

so there never was a Boeing supersonic jet.

Instead, Boeing exited the programme in 1998 when it took over its competitor McDonnell Douglas. In early 1999, therefore, a tight-lipped NASA cancelled the HSCT programme altogether. This also meant the end of the joint research programme with Russia using the Tu-144LL as a flying lab. All of a sudden, the assertions sounded all but encouraging: "The technology we worked on was not robust and wouldn't have carried us to the year 2010," admitted Alan Wilhite from NASA. Boeing's head of the HSCT programme Robert Cuthbertson, declared: "We could have built an HSCT prototype. But economically and technically we found the hurdles too high to build a commercially viable supersonic aircraft. Until we make greater progress in the areas of noise, environmental protection and production it remains unclear if anyone will build a successor to Concorde."

At the same time as NASA and US manufacturers, the Europeans were brainstorming, too, they had first discussed a Concorde successor on the eve of the 20th anniversary of its first flight in 1989. At the time the assumption was it would require US$10 billion. Aérospatiale had already made studies for an Avion de Transport Supersonique Future (ATSF) that was to cruise at Mach 2 to 2.5.

In the UK, British Aerospace was doing research on an Advanced Supersonic Transport (AST). But again, ATSF brochures of the time boast unrealistic predictions for future supersonic passenger markets, the same fantasies the Americans were propagating. From 1994, Deutsche Airbus joined the group with work starting on the European Supersonic Research Programme. The objective was to evaluate the technical and economical feasibility of an aircraft carrying 250 passengers in a three-class cabin over 10,000 km. Again, the Europeans were assuming much overblown potential sales of possibly 500 to even a thousand aircraft. But it seemed that such a vast undertaking would not be possible against the USA, but only if shouldering it jointly with them, and there the Europeans wanted to play an active role and not leave the field to the Americans alone. "At that time there was a general feeling that the future AST would be dominated by Boeing, which was not going to be left behind second-time around," wrote former Concorde test pilot Brian Trubshaw. At the same time he saw Airbus catching up greatly, "which is now a very strong competitor for the No. 1 slot," according to the Concorde veteran in 2000. Still, the Europeans at the time made only a fraction of the effort in their programmes compared to the engagement of NASA and the US manufacturers. Lufthansa engineering director Rolf Stüssel had already declared in 1990: "A transatlantic or transpacific cooperation is inevitable. Because of the limited market, the low number of aircraft and the enormous investments that have to be made for R&D, there can't be competition." In the early 1990s there was even an international working group, of which besides the western European and US companies, the Japanese were also part, as well as Tupolev from Russia. The common studies went very well, according to Brian Trubshaw, although for the whole decade it never came to any concrete projects; still, everyone wanted to be in the loop.

In the Soviet Union and later Russia there had been work on a potential Tu-144 successor since the 1980s and well beyond the dawn of the new millennium. Under the "Tu-244" branding there were several versions of a future SST. Among them was one supposed to fly up to 9500 km fuelled by deep-frozen cryogenic gas. Tests had allegedly yielded excellent results and the gas was supposed to enable cost savings of 70% compared to kerosene. Of course that venture was not launched by accident as Russia boasts the world's biggest reservoir of natural gas and the Gazprom conglomerate

was sponsoring such research. But then it became silent again around utilisation of this exotic fuel. The final design of the Tu-244 as put forward in 1999 was supposed to make a possible first flight between 2005 and 2010. It was basically an enlarged Tu-144 with improvements in wing geometry and engines, this design also boasted a delta wing, but no horizontal stabiliser. The length envisioned was 88 metres, takeoff weight 320 to 350 tons, 250 to 300 passengers were supposed to fly up to 7500 km. As all second-generation SST designs the Tu-244 concept was not fitted with a droop nose, instead there was supposed to be an electronic projection of a live camera feed into the cockpit on takeoff and landing. The joint flight tests of Russians and NASA on board the Tu144LL research aircraft in the late 1990s had yielded important results. Building a next-generation SST had been the pet project of Alexey Tupolev. But Tupolev died in May 2001 at the age of 76. "This was the end of an era. The loss of the man who led Russia's supersonic programme was a big blow to the Tu-244," commented Rob Lewis in his book *Supersonic Secrets*. After the death of the patriarch and during a period of economic and political hardship in Russia the project soon found itself pushed aside.

At the beginning of the new

millennium, after the Russian ambitions had waned and NASA's programme ended, there was a certain disillusionment, as it became obvious that it was not yet technologically possible to build a new SST within parameters that would have found acceptance with governments and the public. While only a few years earlier the future of SSTs had been looking fairly bright, now things had dimmed somewhat. All the excitement of the early 1990s,

when people thought the next-generation SST was imminent, had calmed down a bit, asserted Brian Trubshaw in 2000. Then in March 2001, Boeing put forward a new concept trying to offer passengers higher-speed travel. As one could not penetrate the sound barrier without taking into account the nasty problem of sonic booms, Boeing focused on the new area of Near Supersonic Transport (NST). In other words, transonic flights in the range between about

Mach 0.8 and 1.2. This means there are airflows around the wing profile both of subsonic and supersonic speeds, while the cruise speed of Boeing's proposed *Sonic Cruiser* would have been between Mach 0.95 and 0.98. This would have been 15% faster than other airlines, but only about half of Concorde's speed.

The futuristic-looking aircraft with its curved delta wings, front canards and double tail should have shortened a transatlantic

The oddly shaped Supersonic Boom Demonstrator (SSBD) of NASA, a modified Northrop F5E flew already research missions in 2003 (NASA)

With the single-seat X-59 Quiet Supersonic Transport (QueSST) NASA validate a new low-noise supersonic concept from 2021 (NASA)

flight between Western Europe and America by about 90 minutes. Many observers saw Boeing's concept either as a desperate attempt to take the bull by the horns, but only timidly, or as a distraction from Boeing's problems. Flights over land would have been possible, but the assumption was ticket prices would double – for a time saving of just 12% on average. The transonic zone brings out the worst flight conditions of all, both in aerodynamics and economically. "Nobody should play around near the sound barrier," explained NASA specialist Domenic Maglieri, as this is the worst area for fuel efficiency. "Break the sound barrier and that's it." The transition from supersonic to subsonic airflow in the air stream results in a high flow resistance. The ensuing buffeting of the aircraft shortly before passing the sound barrier was an unpleasant experience already experienced by Chuck Yeager in 1947 on the first supersonic flight. Out of this multitude of negative considerations, the airlines reacted very reservedly to the Sonic Cruiser. They rather wanted an efficient, smaller, long-haul aircraft. Then came 9/11. Boeing quickly buried the *Sonic Cruiser* and came up with an aircraft design that resulted in the successful 787. The end of the story: there still is no more speed in the air.

Now it was the turn of the researchers again, NASA was making use of mathematical findings from the 1960s about the relation between aircraft shape and size for the design of a test plane. The objective was to show that with an appropriate design the sonic boom on the ground could be mitigated. There was rarely a less attractive aircraft than the heavily modified Northrop F5E, morphing into the *Shaped Supersonic Boom Demonstrator* (SSBD). The 1960s fighter received a massively blown-up front section, now resembling the beak of a pelican. In August 2003 first flight-testing was done at Edwards Air Force Base, from exactly where Chuck Yeager had conquered the sound barrier for the first time half a century earlier. The SSBD and a regular F5E flew one after the other over the dry salt lakes with microphones recording the intensity of the boom. The result: the standard boom triggered a shock wave of almost six kilograms per square metre, while the modified SSBD reached only four kilograms, a third less. There were a total of five test flights, during two of them a modified NASA F-15 fighter was following to measure the shock waves of the SSBD. In early 2004 flights were conducted at high altitudes with Mach 1.4. All flights confirmed the scientist's assumptions: an aerodynamically modified supersonic jet also altered its sonic boom. "We can't change the physics of the sonic boom," explained NASA project engineer Ed Hearing afterwards. "We are ploughing through the air faster than it can escape. The solution is to distribute the energy around the aircraft differently to make the result not as noisy."

As the main takeaway from these tests, it became clearer what a future quieter supersonic aircraft would have to look like: extremely sleek in shape, with thin, sharply swept wings. To prevent the shock waves triggered there from reaching the ground, the engine inlets would be mounted above the wings and shaped to prevent vortices from forming. The cockpit would have to be integrated into the fuselage seamlessly, enabling ground vision for the pilots only through a camera feed. At the same time NASA scientists were aware of the fact that, besides minimising the boom, there were many other challenges to solve in parallel; propulsion for example, laminar flow, suitable materials and fuel burn.

The aerodynamicists had a tough problem to solve: it had been shown that designs causing a lower boom at the same time produced increased drag. And it had to be determined which kind of remaining noise of the boom was bearable for the people on the ground. For this purpose, NASA had repeatedly run flight tests with test persons on the ground and precision microphones. Between 1965 and 1967, in the context of preparing for an American SST, there had been a National Sonic Boom Evaluation, a nationwide test programme for sonic boom research. The outcome was that there is a tolerance level of about 70 or 80 decibels, equalling the perceived noise on a busy road. Concorde reached 102 to 105 decibels in supersonic flight, which could be painful for split seconds and damage windows. Due to the logarithmic decibel scale, Concorde's noise surpasses the human tolerance level many times.

At the end of the second decade of the millennium, NASA's activities reached a new level. Now specifically designed aircraft are being built whose only objective is to prove that supersonic flight can be quiet enough to not disturb anyone on the ground, even on overland flights. Back in 2016 NASA handed an assignment to the Lockheed Skunk Works in Palmdale, California, for design and construction of a test aircraft called Quiet Supersonic Transport (QueSST), due to fly in late 2021 or early 2022 for the first time. In preparation, NASA conducted flight tests in 2018 in Galveston on the Gulf of Mexico, an area where the population is not used to sonic booms. A typical sonic boom, visualised in a graphic with the axis of intensity and time, resembles the letter "N". A NASA F/A-18, a research version of the fighter, flew specific manoeuvres breaking the sound barrier coming from subsonic speed in a dive to induce a boom.

Its acoustic footprint was supposed to reach the ground in a way that resembled the noise profile sought with QueSST, more like a sinus curve. The shock wave was supposed to split into several smaller shocks rather than an extremely loud double bang, now only resulting in a short, muffled rumble. While Concorde was still triggering atmospheric overpressure above 95 pascals, the new experimental aircraft was due to cause a shock wave of only just 14 pascals, due to its special shape. And in fact, in the city on the Gulf of Mexico the impact was just heard as "boomlets", muffled boom sounds resembling the closing of a car door from some distance. The researchers assumed that many inhabitants of Galveston might not have even noticed any of it.

It is the idea of the single-seat QueSST demonstrator to reduce both main factors that trigger sonic booms – the aircraft mass and, discovered in the 1950s, the lift. The volume of the aircraft can be reduced if it is distributed over a significantly longer fuselage, so as to have the shock wave form more gradually and be distributed more evenly. The new test aircraft therefore has a long, stretched and pointed front section, also so as to be able to simulate the shock wave signature of significantly larger aircraft. It has received the name X-59 by the US Air Force in 2018

The general shape of NASA's X-59 Quiet SuperSonic Technology airplane, including parts of its wing, can be seen in this overhead view of Lockheed Martin's Skunk Works. [Lockheed Martin via NASA]

and thereby has been admitted into the famous X family of experimental aircraft of NASA and the Air Force. The problem of lift being one of the causes of the sonic boom, however, is hardly possible to resolve, as lift is necessary after all to keep the aircraft in the air. Almost US$250 million of NASA funds have been allocated for the production of the QueSST X-59, in early 2020 assembly of the 29-metre long fuselage and wings with their 9-metre span was already in full swing, by September the aircraft was half way assembled. With

these dimensions the X-59 is almost five metres shorter than a Boeing 737-700, with a wingspan, however, that is two metres shorter than that of a single-engine Cessna 172. The short wingspan reduces drag, while the delta wings still deliver sufficient lift. Cockpit, ejection seat and cockpit canopy have been taken from Northrop's T-38, the gear comes from the F-16.

Thrust for the X-59 is generated by a single GE F414 engine used in the F/A-18 Hornet, with the test aircraft supposed to reach Mach 1.5 (1590 km/h) and fly at

altitudes of up to 16,800 metres. The central engine intake above the wing also works to minimise noise as it is shielded by the wing from reaching the ground. The air intake, however, could be disturbed by vortices emerging from this configuration. The aircraft noise perceived on the ground is supposed to be around 60 db(A), about a thousandth of what current military supersonic jets can cause in sound emission. The canards in front of the stretched fuselage are also contributing, as they hinder a merger of shock waves. The long and fixed nose prevents ground and forward vision for the pilot, therefore for the first time an Enhanced Flight Vision System (EVS) is installed with a forward facing 4K camera delivering a live feed for the cockpit. As less weight also reduces noise, there is no doubt that future supersonic aircraft have to make do without heavy droop nose mechanisms as in earlier designs.

But then the X-59 is of course not a passenger aircraft, it does not transport anything, its only task is to prove the sonic boom can be lowered to an acceptable level. NASA also calls it "Boom Box". Dominic Maglierei, NASA expert for supersonic flying, says: "It is supposed to climb

SUBSONIC FLIGHT

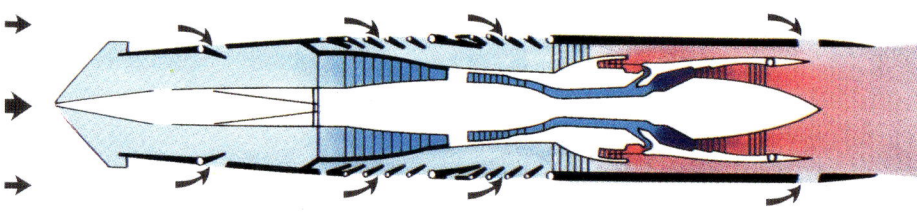

SUPERSONIC CRUISE

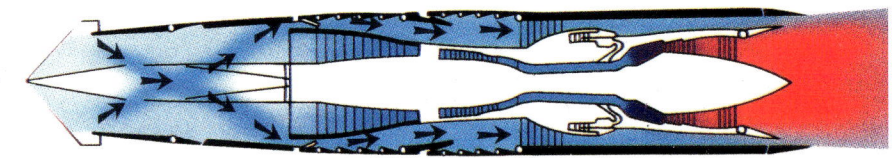

A decisive factor in engine layout for supersonic jets is their adaptability for different modes flying subsonic or supersonic (AS)

to 55,000 feet, get into cruise mode and produce a 75 PLdB footprint. That's all." During 2023 flight tests are due to begin and last until 2025. The perception of supersonic flight and sonic boom on the ground will be measured in different parts of the US, inside and outdoors, as well as its strength and variety. Three times a day or night the aircraft will perform two overflights of almost 100 km each, within 20 minutes of a certain town, in different weather conditions, under instrument flight rules and hopefully with sustainable fuels. And why all this fuss? To finally convince the FAA and the world aviation organisation ICAO to lift, or at least loosen, the ban

on supersonic flights over land in place for almost 50 years. This would be a very welcome shot in the arm for the different new (and so far exclusively American) companies bringing new supersonic aircraft projects to market, even though their concepts so far don't rely on it, for good reason.

It is a fact that there is currently a new momentum in terms of supersonic flying. It is based on the advances of new technology now being available, as well as the political will in the US to stay abreast of these changes and take a long hard look at the antiquated rules, that at the time were mostly politically motivated and directed solely at Concorde. In December 2019, the FAA announced: "Lighter and more efficient composite materials, combined with new engine and airframe designs, may offer the potential for introduction of a viable SST. … The Department of Transport (DOT) and the FAA are taking steps to advance the development of civil supersonic aircraft." One of them is a new rule about the noise of such aircraft, another one an easing of rules to get certified for supersonic test flights over US territory. For this, according to the FAA, there is also support in US Congress. While the proposed activities to define new rules would not equal a lift of the ban to fly above Mach 1 over land yet, the authority pledged nevertheless to rethink

the spectrum of supersonic operations that might be permissible in the future. Starting in December 2020 there will be a bi-annual evaluation of aircraft noise data to determine if lifting the ban might become possible.

In January 2021, the FAA announced that applications for overland test flights are now possible on a case-by-case basis and clarified the rules. Before permission might be given, there will be a thorough evaluation each time of both the manufacturer's case as well as the test area's application, mostly determining the expected environmental impact. "The FAA supports the new development of supersonic aircraft as long as safety parameters are followed," said FAA Administrator Steve Dickson. "The testing of supersonic aircraft at Mach 1 will only be conducted following consideration of any impact to the environment." This was welcome news to the new players. The CEO of Aerion, Tom Vice, hailed it as a "significant milestone in the development of civil supersonic flight", he is thankful to now have "the ability to test the AS2 aircraft over land in addition to overwater testing currently planned".

And almost instinctively, the opposite side came out of hibernation, as in the early 1970s it immediately showed how emotionally charged the subject of supersonic flights

still is in the political discussion. "With the existing technology there is no way to advance the 'safe and efficient operation of supersonic aircraft'", 28 scientists and environmentalists declared in a letter to the FAA. In which they ignore the fact that no one intends to do this with existing technology. "It is expected that new commercial supersonic aircraft will burn five to seven times as much fuel per passenger as comparable subsonic aircraft, and that they will exceed international limits for CO_2 emissions by 70%. To allow a new class of super-polluting aircraft in the sky would be madness and obviously not match the FAA's obligation to protect public safety and wellbeing," they go on in their letter, culminating in their demand to the authority: "Don't support the revival of this failed and destructive technology." In a direct reaction, the new players Boom Aerospace and Aerion Corporation, the private enterprises pushing new supersonic aircraft concepts, almost simultaneously adapted their strategies with both now stressing that their future aircraft will exclusively fly 100% with sustainable aviation fuels.

In June 2019 Lockheed Martin made news again while still producing X-59 for NASA: the company introduced a concept study for a low boom SST called *Quiet Supersonic Technology Airliner* (QSTA). Even at first

The Quiet Supersonic Technology Airliner (QSTA) unveiled by Lockheed Martin in 2019 is supposed to carry up to 40 passengers with Mach 1.6 over water and 1.8 over water (Lockheed Martin)

sight it is obvious this design encompasses all those technologies and constructive approaches tried out on a smaller scale in the X-59 already.

The 69-metre long delta wing aircraft with a span of 22 metres could carry 40 passengers for up to 9360 km. This makes the QSTA seven metres longer than Concorde, as its nose section is extremely stretched for aerodynamic reasons, while its wingspan is about three metres shorter. The two turbofan engines are integrated into the unusual double tail in V-form. There would be no afterburners, but the propulsion system would have to be developed almost from scratch to deliver the required 40,000 lbs (178 kN) ground thrust. Range of the QSTA would be sufficient for transatlantic routes and even for

Lockheed Martin's first draft of a low-noise supersonic airliner still looked slightly outlandish (NASA)

Supersonic Travel

nonstop transpacific flights, for example from Los Angeles to Tokyo. That is only achievable with overland flights above Mach 1, possible due to the much lower sonic boom impact. With a planned cruise speed of Mach 1.8 over water the QSTA would be slightly slower than Concorde. It would then even be viable for routes from Europe to Asia like London to Tokyo. This city pair would be at the upper end of its range, but the QSTA could deliver time savings of four and a half hours versus a subsonic jet. Over land there would be a limitation to Mach 1.6 in any case due to noise abatement. There is also a whole package of innovations physically enabling a new, low-noise supersonic era: smaller aircraft with less mass and aerodynamically engineered to minimise the sonic boom. Their quieter engines featuring medium bypass ratio and no afterburners, with the upward-facing inlets placed on top of the wings to deflect shock waves. The engines work with laminar flows, enabling greater range and reducing emissions. Also omitting a droop nose with heavy hinge mechanism in favour of a digital lookout for pilots. That sounded so convincing to some experts that they claimed the QSTA was the clearest sign so far "that we are standing at the edge of a new golden age of super fast travel". But first it has to be proven by the X-59 that supersonic flying in the 21st century can be done sustainably and with low noise emissions. "Ending this ban would create possibilities for a whole new industry," according to Lockheed Martin. The opportunity to go anywhere in the world in half the time it takes today would open up "a huge potential market for aircraft manufacturers to develop quiet commercial aircraft with the insights we have gained with the X-59". If this will be achieved, Lockheed Martin wants to have a concept in hand, the QSTA, to take advantage of new commercial routes with an SST. It has been conceived in over two decades of research work. And it has been, and still will be, a long way to go.

Boeing's newest study for a civil supersonic airliner resembles other current research projects (Boeing)

thirteen

Supersonic business jets: AS2 and Spike S-512

"While supersonic aircraft might be ushered in by the wealthy, they will ultimately benefit anyone who wants and needs to get to their destinations faster."

Vik Kachoria, CEO Spike Aerospace, 2019

Aerion AS2. [Aerion]

Large supersonic airliners of the size of Concorde or Tu-144 cause great problems in terms of environmental compatibility and economical viability. So, it is tempting to lower the benchmark for the next generation of SSTs and start by trying one size smaller. All designs put forward so far for supersonic airliners of the future assume they will have less mass than Concorde. Only by doing this is there the chance to lower engine noise on takeoff and landing as well as the sonic boom footprint enough to have the aircraft find environmental acceptance. As it is the objective of manufacturers to firstly get a grip on technology in order to build a more sustainable and lower-impact supersonic aircraft, focusing initially on a specific clientele seems to make sense: people for whom speed is paramount and who are willing and able to pay for it. These are the super-rich, who dominate the market for large business jets. "It all comes down to the ability to 'create time'. Time is one of the rarest commodities we have. What our airplane can do is convert non-productive to productive time, whether it's for business or leisure. Essentially (our aircraft) the AS2 has the utility of a time machine," a former CEO of Aerion Supersonic said in 2015. "The AS2 will be practical, efficient and provide a value proposition that is tremendous in terms of saving time." That was the plan – until Aerion abruptly ceased operations in May 2021, citing challenges to close on large new capital requirements, estimated to be at more than US$4bn, to establish AS2 production in Florida as planned. Nominally Aerion had an order backlog worth over US$11bn, mostly from fractional ownership firms like Flexjet and NetJets, but it was seen as fragile, meaning no money had changed hands as yet.

Let's look back at how it all evolved. Aerion estimated that there are about 3600 billionaires, and the number increases by 8% annually. More and more of them, however, hail from China, a country with very harsh restrictions on the use of private aircraft. And the majority of billionaires come from the US, where there still is a ban on supersonic flying overland in place as described earlier. Still, since the 1980s it became more and more likely that the next civil supersonic aircraft might be a business jet and not a large airliner. In the field of smaller supersonic jets there has been substantial progress in developing such concepts. This has been driven by the engagement of US billionaires, who found developing private supersonic business jets a tempting and potentially lucrative field of work.

Back in 1989, the first project for a Supersonic Business Jet (SSBJ) had already begun. The leading US business jet manufacturer Gulfstream had teamed up with Soviet design bureau Sukhoi to develop the concept of the S-21. At the time, the assumption was that development of an SSBJ would cost between US$1.5 and 3.5 billion and such an aircraft would sell for between US$80 to 100 million apiece. Market volume for such a small supersonic jet was projected to be about 300 aircraft, about the same potential as a successful business jet. The S-21 was to be a trijet with six to ten seats, it was supposed to fly for up to 4300 km at Mach 1.4, or even up to 7400 km at just Mach 0.95. The plan was that Sukhoi should build the fuselage and wings, while Gulfstream would integrate engines and avionics. But the project did not last long and ended in 1992. Gulfstream was taken over by a new majority owner with different priorities, the market outlook was uncertain and the troubles of the post-Soviet era had left its mark on Sukhoi. In other places, concept drawings and models of three-engine SSBJs surfaced as well, or at least announcements were made. French business jet manufacturer Dassault started evaluating a "Falcon SST", a supersonic iteration of its Falcon 50 with a similar cabin. It was to be equipped with three engines developed from existing Western

British Aerospace put out a first draft of a Supersonic Business Jet (SSBJ) in the early 1990s (AS)

Already back in 1989, Gulfstream and Sukhoi cooperated on a joint project for an SSBJ (AS)

F-104 Starfighter with its un-swept, stubby wings. During modern supersonic research, NASA started with a variety of flights testing laminar flows in 1995, initially with a modified F-16XL fighter aircraft, then in 2000 with an F-15B. The assumptions were confirmed in these tests and NASA attested that laminar flows indeed gave supersonic jets the potential to achieve a similar, or in parts even higher, efficiency than subsonic aircraft on a comparable mission.

The potential for supersonic flying identified in laminar flows was one of the reasons that led to the establishment of Aerion Supersonic in 2002, initially based in Reno/Nevada, even before Concorde flights ended. Among the founders was aerodynamic pioneer Richard Tracy, who became Aerion's Chief Technology Officer, while the money, until the sudden end of the company in 2021, came from Texan oil billionaire Robert Bass. In decades of research, Tracy had developed the first wing for a supersonic aircraft with natural laminar flow. There are only a few aircraft in history which were capable of supersonic cruise without afterburners, besides Concorde; these were the Lockheed SR-71 Blackbird and the Convair B-58. They all suffered from the problem of extremely high

ones, without afterburners, and fly at Mach 1.8. Also, British Aerospace put out visuals of an SSBJ design featuring canard wings in the front above the cockpit, and three engines as well, two mounted on top of the wings and one in the root of the vertical stabiliser. With this projected business jet, twelve passengers would have flown up to 7000 km at Mach 1.85. But in the end, these all were vague announcements and declarations. At the time there was no solution as to how to mitigate the sonic boom and how to develop engines with a sufficiently long life; changing priorities at the manufacturers

also soon pushed these projects aside.

But, behind closed doors research went on, as getting a grip on muffling the sonic boom was only one of the challenges. The other was the large drag in cruise flight. To lower it, and thus increase range, speed and fuel efficiency, using laminar flows seemed promising. The hope was that a straight wing surrounded by laminar flows would be beneficial despite the wave impedance, as this configuration substantially lowers friction drag. The first aircraft that followed this design philosophy back in the 1950s was the interceptor Lockheed

The S-21G from Gulfstream and Sukhoi was designed as a six-to-ten seater for Mach 1.4 (AS)

Early draft of an Aerion SSBJ in 2004 (AS)

fuel consumption, and that could be traced back to their wing shape. Every surface in an air stream experiences a laminar boundary layer, from a certain length this turns into turbulence. That happens earlier the more friction is created on the wing surface, and the more transverse flows exist in the air stream. Laminar flows cause substantially less air drag than turbulent streams. The non-swept wing with its beneficial flow characteristics therefore became the core concept of the SSBJ conceived by Aerion, first unveiled in 2004. The wing design was supposed to hinder air streaming outward over the wing, resulting in less wing-induced drag. The aerodynamic patent of Richard Tracy was supposed to enable the desired

laminar flow over 70% of the upper wing surface and up to 100% on the underside.

This seemed to enable the use of two just slightly modified Pratt & WhitneyJT8D-219 engines, also found on the Boeing MD-80. In 2004, the expectation was development costs would be around US$1.2 to 1.4 billion and the billionaire's toy, costing US80 million apiece, would be on the market by 2011. The fuselage was supposed to be manufactured from conventional aluminium, the extra thin wings from composite material, with only their tips made of titanium for heat resistance. While the cabin width narrows towards the back adhering to the area rule, the lower mid-segment has strakes sticking out to the sides, containing both main gear and

fuel tanks as well as supposedly leading to aerodynamic improvement. Aerion estimated demand for 250-300 aircraft in the first decade, of which up to half could be by fractional ownership companies, with users buying shares and flight hours, but not the aircraft.

But uncertainties and delays of a few or even many years are apparently unavoidable, a given for almost all civil aircraft programmes and seemingly multiplying in unknown territory such as developing modern supersonic jets. In the case of Aerion this added up to almost two decades passing since the company started and when it ceased operations unexpectedly in 2021, still no supersonic jet had taken off, the name having now been changed to AS2.

Time and again Aerion took apparently strong partners on board, initially teaming up with Airbus, only later to have them decide such a niche market wouldn't fit the portfolio. Aerion

then partnered with Lockheed Martin, which also only lasted for a year. Since 2019 Boeing had taken a share in Aerion, giving it credibility and standing. That's one of the reasons the industry seemed to agree: if anyone would have been capable of bringing the first supersonic passenger jet to market in over 50 years, the first next-generation SST, then it would have been Aerion. As it turned out, this was a misjudgement, once again proving the volatility of any supersonic undertaking and the giant obstacles it faces, no matter how potent a company appears to be.

Aerion had proven to be surprisingly resilient, facing a multitude of hurdles that had led to the demise of almost all other companies aspiring to restart supersonic flight – and ultimately its own. Aerion was seen as a

Aerodynamics pioneer Richard Tracy with a wind tunnel model of the Aerion design (Aerion)

Only in 2020 Aerion switched from stub to delta wings on its three-engine AS2 (Aerion)

"start-up veteran" sticking to its original dream, while it initially had been haunted by bad timing when the global financial crisis hit, paralysing the business jet segment between 2008 and 2011. By the time it was ready to start over, noise regulations at airports had been tightened in a way that precluded the intended JT8D engines from being used. That made it necessary to configure the AS2 as a trijet even before alternative propulsion was chosen, with the middle engine embedded in the root of the vertical stabiliser and the air inlet located aft above the cabin. In the meantime, the list price per aircraft had risen to US$120 million.

From 2017, noise rules for aircraft above takeoff weights of over 54,400 kg were tightened again. Therefore, the strict objective was to keep the design below this threshold by any means, although every engineer knows that aircraft always gain weight during development. Also, market research showed that the customer focus had shifted away from transatlantic routes, with transpacific now being more in demand. All these changing conditions brought several design and specification changes of the AS2 over the years. The original 2004 design was hardly recognisable in its final iteration after radical changes in 2020. The cruise speed had been lowered to Mach 1.4, while boosting range to 8700 km so it was capable of flying transpacific routes. In the meantime, there was repeated silence regarding the project, before news emerged again. In 2017 the project got an important boost, literally, when the most crucial decision was taken, the point at which other supersonic projects have failed before: General Electric (GE) announced it was developing an engine for Aerion called Affinity, which was supposed to become the first new supersonic propulsion in more than five decades. That programme was also discontinued following the demise of Aerion in May 2021. The digitally controlled high performance turbofan was not entirely new, also in order to contain costs that otherwise could have been over US$1 billion. Instead, the core of the widely used and reliable CFM56 was to be used as the base, the engine that provides propulsion to medium-haul workhorses such as the Boeing 737 or the A320 family. A new twin fan low-pressure section was being developed to adapt the engine, allowing efficient cruise, both supersonic over water as well as subsonic over land. Compared to the Olympus engines of Concorde, Affinity wouldn't have needed an afterburner, neither on takeoff nor to accelerate or during supersonic cruise,

The original draft of the Aerion AS2 called for stub wings similar to the Lockheed F104 Starfighter (AS)

with the bypass ratio being the highest ever in a supersonic engine, according to GE. The company has longstanding experience with such propulsion systems reaching back to the 1950s Lockheed F-104 Starfighter, whose J79 engine was also built by GE.

With Affinity, GE also wanted to invest in potential supersonic airliners of the future, as Aerion was not hiding its intention of using the AS2 just as a base model. "Aerion has a 50-year technology roadmap. Our first step toward this faster future is with the supersonic AS2," Aerion's CFO and EVP Mike Mancini said in 2019. The next step could be an airliner. "We want to do this thoughtfully and in an environmentally responsible way so that we avoid all of the problems that limited the

potential of the Concorde many years ago," stated Mancini. "It would be reasonable to expect a first generation of supersonic airliners in the 2030s. Buoyed by such bold plans, the old mentality of the industry to predict alleged high potential market volumes popped up yet again. According to an independent 2016 study, by 2035 there could have been sales of up to 1300 supersonic aircraft. How much caution one has to take with such projections has been described at length in this book already. While the objective was to adhere to newest chapter 5 noise standards at takeoff, Aerion was also boasting of another advantage it called "boomless cruise". Behind it is a physical phenomenon, known to experts as Mach cut-off. Roughly, this can be described

as limiting the boom by limiting supersonic speed. The objective is to be able to fly overland at about Mach 1.1 to 1.2, without any boom reaching the ground. The speed of sound mostly depends on the air temperature, and the Mach cut-off makes use of the different air temperatures between cruise altitude and on the ground in a way that the boom dissipates on its way down. To achieve this effect, maintaining a certain altitude is necessary, from which the conditions are exactly right to take advantage of them. "At speeds up to about Mach 1.2, the AS2 will deflect the boom upwards off warmer layers of the atmosphere and therefore it will not be heard on the ground," explained Mancini. For the cockpit, jointly designed with Honeywell, a technology

was developed to measure atmospheric conditions in direction of the flight to enable the aircraft to strictly adhere to the exact speed making the effect possible. "We see this as the only practical boom technology for years to come, if not decades. We have trademarked this capability as 'Boomless Cruise'," stated Mancini. It appeared to the Aerion engineers that creating the aerodynamic conditions for a low boom through the aircraft's design was not economically viable.

In today's context of discussing flight shaming and insistence on sustainability in the behaviour of every individual, the question is: do supersonic flights for a small fraction of elitist passengers make sense,

as they substantially burden the environment? Initially, the scenario might sound tempting: business travellers, let's say from Germany having to go to New York, are on the road for two or even three days. With an SSBJ this would look completely different: takeoff in Munich at 8am local time, flight time to New York about four hours, arrival at 7am local time. Including all ground and transfer times a meeting in Manhattan could start at 9am. Very time-pressed individuals could go back to Europe at local lunchtime, as long as they are allowed to land there close to midnight local time, which most airports don't allow. Or, one spends a whole day in Manhattan, goes out to the airport around 6pm, takes off

Draft for the Spike S-512 SSBJ (Spike)

from New York around 8.30pm and comes in to land after four hours of mostly supersonic flight before 7am the next morning in Munich. Then to catch up with sleep, as not every tight business jet offers the sleeper seats big airlines have, and finally maybe work for half a day. Is that desirable? And for what price, beyond the financial sum, but also incorporating the strains to the environment caused?

In early 2019, the International Council on Clean Transportation (ICCT), an international panel of scientists, put out a study about the anticipated noise and climate impact of an "unlimited commercial supersonic network". The scientists take an anticipated scenario in 2035 as their base, in which 2000 supersonic jets connect 500 city pairs with about 5000 daily flights. The two most utilised airports would be Dubai and London-Heathrow each with over 300 supersonic movements a day. Big parts of the world would be exposed to the sonic boom as a disturbance, in countries as centrally located as Germany, this could be the case every five minutes for up to 16 hours daily, 150 to 200 times a day, according to this scenario. At the same time, the scientists assume that a typical supersonic airliner would exceed all maximum levels currently valid for harmful nitrogen oxide emissions by 40% and for CO_2

by as much as 70%, as well as surpassing all noise limits at airports. Fuel consumption would be five to seven times as high as for subsonic aircraft. Compared to modern current jetliners, fuel consumption of an SST would be three times higher than for current Business Class passengers and as much as nine times above today's fuel burn per Economy passenger. The main assumptions of the study appear to be overly negative in respect of the current supersonic projects. Still, there is a valid question it raises: how about the social acceptance of a new supersonic era in the 21st century? "While the public were, on the whole, very acceptant of the environmental externalities of Concorde, the reaction to the noise and pollution made by business or rich people-only SSTs might not be so tolerant," assumes the Royal Aeronautical Society.

It did not take the manufacturers long to answer. Vik Kachoria, the CEO of Spike Aerospace from Boston, still focused on the future market of smaller supersonic jets even after Aerion's demise, rebutted straight away, on the day after the study was released. He pointed out unfounded and invalid assumptions by the scientists and a denial by the ICCT of technological progress made. "Most of Spike's engineers, management team, partners and customers have

families and children too. We are not out to damage the environment and are out to ensure our jets do not cause the scenario they describe. Instead, they assume we have no interest in the planet or the environment and just want to destroy it with a toy for the rich," said Kachoria. He stressed that an SSBJ developed by Spike would not cause a noisy sonic boom. "It is completely unacceptable to advance technology or transportation at the detriment of the environment or the community. That is simply irresponsible." Aerion reacted as well: "We are aiming at every AS2 flight from day one being CO_2 neutral," explained CEO Tom Vice. "We have worked on an engine with GE that offers great progress in efficiency and is designed from scratch to run on 100% sustainable aviation fuel" – not with just a 30% drop-in into conventional fuel as is common today. That, stressed Vice, would result immediately in a net reduction of CO_2 by 80%. Aerion also announced they would plant a hundred million trees and, additionally, would pay for CO_2 compensation for their customers. The green conscience of an industry that is not green on first sight had definitely woken up.

After there had not been any news updates on Aerion's

Final configuration of the Aerion AS2 with delta wings (Aerion)

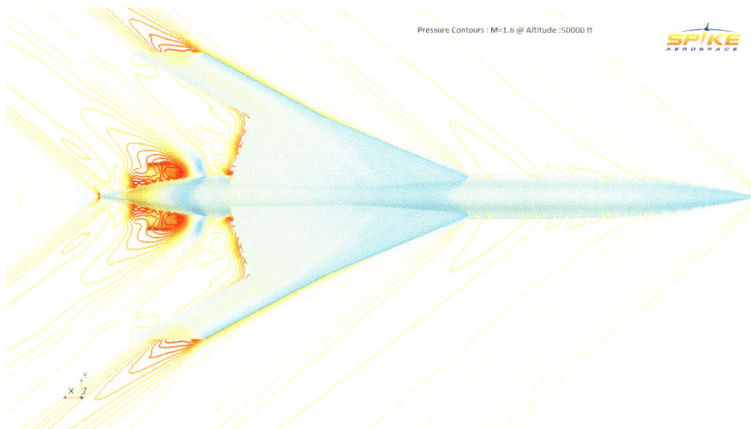

Pressure Contours : M=1.6 @ Altitude :50000 ft

This graphic shows the shock waves of the Spike S-512 when flying at Mach 1.6 in about 15,000 meters of altitude (Spike)

aircraft design for a long time, the company presented a radically altered concept in April 2020. And, all of a sudden, the AS2 re-appeared with delta wings, as basically all supersonic aircraft ever before, replacing the straight stub wings that had been considered Aerion's trademark all along. On the way, the wingspan had been increased by about half a metre to 24.07 metres, while the fuselage came out six metres shorter at 44.17 metres. Now there was much less natural supersonic laminar flow on the aircraft, as the different objectives pulled the design in many directions at once. "You see a lot less supersonic natural laminar flow on this airplane because we had to get an aircraft that could meet all of the different competing flight regimes. Our task was to optimize around those competing requirements

and come up with an aircraft that could meet them all simultaneously," said Aerion CEO Tim Vice. Presumably under the influence of Boeing, Aerion had decided that a delta wing was more beneficial to flight physics, aircraft length and drag. A major design change affects also the outlet of the middle engine, which is now lead underneath the vertical stabiliser, while before it was led straight through its root, resulting in the whole tail section now becoming smaller and the aircraft significantly shorter. This new configuration also would have allowed for a lower angle of attack on landing compared to Concorde, no droop nose required, but the pilots would still have gotten an Enhanced Vision System with the projection from an outside camera fed into the cockpit to enhance spatial orientation.

The maximum takeoff weight

had increased from 54.8 tons before to now 63 tons. This meant that in the redesign, almost 32 tons of fuel could be carried, compared to just 28 tons before. Similar to Concorde, the centre of gravity was to be optimised during cruise through a complex fuel transfer to trim tanks. But the AS2 would still not have been capable of flying non-stop from New York to Tokyo (distance about 10,800 km), but would have been required to put in a fuel stop, narrowing the precious time gain. The newest subsonic business jet, Gulfstream G700 with a cabin of similar size, in contrast, has a range of about 13,800 km and flies fast and efficiently non-stop. Still, experiences of business jet manufacturers show that only few jet operators use up almost

all of their range to the limit. "We really wanted to have an aircraft that was superefficient at Mach 0.95 and superefficient at Mach 1.4, while at the same time minimizing noise," said Tom Vice. "These are not loud aircraft. In fact they meet the most stringent noise rules." The FAA would have probably accepted a maximum over-flight noise measuring 94 EPNdB. At the time of the supposedly final concept change to the AS2, first flight seemed possible in 2024; now this will not happen at all. "Concorde was a bold, noble experiment and a major milestone in the history of aircraft, but the AS2 is very different in both specification and business model," said Aerion CEO Tom Vice. "We'll have an economically viable production volume of 300 units

in our first ten years – plus the backing of several strong investors." Before shutting down, Aerion was aiming for the AS2 entering service in 2026. In March 2021 Aerion for the first time revealed a glimpse of its further, even bolder plans: a commercial airliner called AS3 travelling at speeds of "Mach 4+", according to Aerion, carrying 50 passengers over 13,000 km. It is "targeted to take to the skies before the end of the decade and build upon the AS2 business jet to bring Aerion's innovation to commercial air transportation," the company said. While Aerion was known to pursue revolutionary concepts out of the public eye, announcing commercial transport with speeds of about 5,000 km/h within a decade sounded no less than outlandish

The configuration of the Spike S-512 has only been defined in rough outlines (Spike)

The Spike S-512 is the first passenger airliner configured without windows, replace by surround-screens inside the cabin (Spike)

to get supersonic to work first."

Currently known competitors in the market of supersonic business jets are Spike Aerospace from Boston and Virgin Galactic, having revealed a 19-seat delta wing design in 2020. Also, the two US tech companies Exosonic and Hermeus as well as Boom (see next chapter) have won funding from the US Air Force to come up with designs for supersonic concepts. They could be used to redeploy missiles to forward bases in a conflict – or used as a new supersonic variant of Air Force One. While little details are known, Spike has

been around the longest. Since 2013 the company has been conceiving an SSBJ named S-512. It will be larger than the AS2 with a length of 37 metres, a wingspan of the delta wing at 17.7 metres and the cabin seating up to 18 passengers, at some point Spike even spoke of 22 seats. The aircraft will have two tail-mounted engines delivering a thrust of 88.9 kN each, reaching Mach 1.6, faster than the AS2. A time gain of 40 to 50% versus a subsonic jet is what the manufacturer promises. Maximum range is planned to be 11,400 km, enabling the S-512 to fly routes

at this time. This would have meant an airliner would finally reach the low hypersonic region, which starts at Mach 5. The fastest aircraft with air-breathing engines so far is the military surveillance plane Lockheed SR71 Blackbird, which attained Mach 3.3 (about 4,000 km/h). This plane once made it from New York to London in just under two hours on a record flight with air-to-air refueling. Aerion had teamed up with NASA's Langley Research Center to study the future of commercial flight in the Mach 3-5 range. According to Aerion's CEO Tom Vice, there is an interesting sweet spot around Mach 4.5, enabling flying from the US to Japan in two hours or less while avoiding certain challenges with materials and

cooling. Details about the AS3 were vague, apparently its design was to include swept delta wings, twin vertical tails and four engines mounted under the wings. Aerion's objective was to connect any two points on the planet within three hours. "Supersonic flight is the starting point. To truly revolutionize global mobility as we know it today, we must push the boundaries of what is possible," said Vice. Scientist Bernd Liebhardt from the German Aerospace Center. DLR was much more reserved. "For me, hypersonic lies several decades in the future. Supersonic is already a difficult undertaking, and hypersonic is going still another step further," he said. "We need

The Multiplex Digital Cabin can show surround projections of a live camera feed of the outside as well as virtual projections (Spike)

In contrast to almost all other drafts the S-512 is designed as a twinjet (Spike)

such as London to Hong Kong or Dubai to New York non-stop. But Spike specifically focuses on lucrative transcontinental business jet routes for example in Asia, which are only possible by flying supersonic overland, as is part of its business plan, in contrast to the other projects. "Now you can do Singapore-Tokyo, Singapore-Sydney as a day business trip," promised John Thomas of Spike's executive team. "Dubai to London is also a day trip, to me that's a game changer. Because

if you can do that –and not at a first-class fare but a premium-to-business class- the number of people who are going to do that becomes a really interesting opportunity." Spike wants to register a patent for a *Quiet Supersonic Flight Technology,* without so far further giving details, while it wants to work within the same framework as the Aerion venture had aimed for. Spike promises minimal boom impact on the ground not exceeding 75 PLdB, NASA's target for future approval of

overland supersonic flying. Operating costs are supposed to be only 15% above those of a subsonic business jet.

There is another major difference in the S-512: as the first aircraft design ever that has a windowless cabin, replaced by an innovative *Multiplex Digital Cabin.* Which means all the interior walls will be fitted with displays showing either a 360° live panorama shots by the outside cameras, but also entertainment programmes or company presentations.

Windowless aircraft with just virtual lookouts have been a longstanding desire of many aircraft engineers, as it enables them to design the airframe structure to be much stronger and also to save weight. Even back when designing Concorde, this would have been the engineer's way to do it. As at the time, there were no ways of providing virtual views outside, the miniature windows remained as a compromise. Not much is known about technical details and programme progress at

Spike. According to industry media, the company has still much to prove, not least to come up with funding. The only milestone made public happened on October 7th, 2017. At the time, there were successful flight tests with an unmanned, smaller-scale demonstration aircraft called SX-1.2 in New England on the US East coast. Seven short test flights were done with the remote-controlled aircraft on that day, proving the design was "valid", as Spike declared. "The SX-1.2 test flights were concluded in a real world situation, and provide significantly more data than wind tunnel tests done in an artificial environment," explained CEO Vik Kachoria. In 2017 the company had announced that preliminary work on the next demonstrator SX-1.3 had already started. But nothing else about it was ever published, also there were never any visuals of SX-1.2.

There was even talk of "manned high speed demonstrators" supposedly flying back in 2017, with actual flights estimated in 2019, also reports about Spike's plans for a 40-50-seater being in an early stage. The aircraft would create actual demand, not only satisfy existing demand, the chief advisor to Spike declared then. In spring 2021, CEO Vik Kachoria stated that he is "90% confident" in being able to achieve his low-boom targets after now having reached the sixth conceptual design iteration. At the time, wind-tunnel testing was planned for the fall of 2021, but full confidence to reach the noise targets of strictest Chapter 5 levels can only achieved through the long-overdue manned proof-of-concept aircraft, that was now supposedly ready to fly in 2022. Even the CEO admits his plans for the Spike S-512 to enter service in 2028 are "optimistic". So far, Spike has set its list price for the aircraft at US$125 million, positioning its base model already above the AS2. The scepticism of a scientific report in 2017 might be warranted, which said: "The business and technical details proposed by Spike's representative are quite ambitious, particularly compared to those of other companies in the area, which are in general pursuing more incremental technologies with more experienced teams. … This timeline is especially optimistic and will be challenging to achieve." Which has been proven true ever since and probably can easily be applied to the whole market segment.

Meanwhile Russia has not given up its ambitions in the arena of civil supersonic flight. After the end in 1992 of the S-21 project of Gulfstream and Suchoi for an SSBJ and

a bigger design, the S-51 for 68 passengers, the Russian manufacturer kept on working in this segment. From 2005 to 2009 Sukhoi was the Russian partner in the international research programme HISAC (High Speed Aircraft), since then the manufacturer has been cooperating with the famed Moscow TsAGI (Central Aerohydrodynamic Institute). Between 2017 and 2019, the institute received the equivalent of US$20 million for research on a new SSBJ project, in 2019

another US$2 million were allocated for the next step. According to the tender of the Russian ministry of trade and industry, the task put out was to find a compromise between acceptable environmental standards (in sonic boom and on takeoff) as well as define competitive technical and economical standards for a new supersonic aircraft.

In early 2020, a configuration the institute came up with received a patent. It would be capable of lowering the

boom on the ground to 65dBA while the aircraft would fly at Mach 1.7 to Mach 2 in 14,000 to 16,000 metres of altitude. A MIG-29 fighter aircraft flying in about 11,000 metres of altitude at Mach 1.75 causes a boom measuring 92dBA on the ground. Two aircraft versions have been envisioned by TsAGI: an SSBJ for eight to ten passengers, takeoff weight 55 tons, with two turbofans of about 15 tons (147 kN) thrust each, this stands as the concept with a bigger

chance to become reality. Also, a supersonic airliner with the same aerodynamic concept from TsAGI research could result from the efforts. It would be equipped with four instead of two engines of the same type. Weight would be 120 tons, up to 82 passengers could be carried over about 8300 km at Mach 1.8. Chances this will ever happen can, however, be seen as minimal.

For SSBJs once again New York will become a major destination. If the Spike S-512 will ever fly is written in the stars, literally (Spike)

fourteen

Boom:
From the XB-1 to the Overture supersonic airliner

"My aspiration is to make the fastest flight also the most affordable and ultimately get to a place where you can go anywhere in the world in four hours for a hundred bucks."

Blake Scholl, CEO Boom Supersonic, 2021

Boom's XB-1 test bed (Boom Supersonic)

It will be named *Overture* and come from Denver, Colorado, on the edge of the Rocky Mountains in the West of the US. It is supposed to weigh much less than Concorde, up to one and a half times less, be ten metres shorter and to come with a delta wing of seven metres less span, but with a new shape. It will cruise ten per cent faster than Concorde at Mach 2.2 and cover the distance from New York to London in just three and a half hours or even three and a quarter hours. And it will be significantly quieter than Concorde on takeoff, some claim up to 25-30% less noisy. Depending on cabin configuration, between 55 and 75 passengers will be on board and are supposed to pay only as much as a subsonic business class ticket on a transatlantic route costs today. That was US$5000 on average in 2018 for the round trip, while the same journey on Concorde was up to US$13,500 back in 2003. And *Overture* is supposed to become the supersonic airliner of the 21st century, powered by 100% sustainable aviation fuel. At least that is what the boyish-looking founder of Boom Technology Inc. Supersonic plans, engineer Blake Scholl, hailing from Cincinatti in the US Midwest. That sounds bold, and it definitely is. But this is not a pipe dream or fantasy of an overly confident start-up entrepreneur. It is serious enough that private investors had backed the company with funds of US$210 million by the end of 2020. And Japan Airlines not only already invested US$10 million in the company in 2017, but also took an active role in the development and opted for 20 *Overture* aircraft, making non-refundable down-payments. Stunning the aviation industry, which had just tried to come to grips with what the sudden demise of Aerion Supersonic in May 2021 (see previous chapter) would mean for the supersonic airliner market in general and for other start-ups, an announcement was made in early June 2021.

Just two weeks after Aerion folded, United Airlines, one of the world's biggest airlines, announced its intention of buying 15 *Overture* and signing options for an additional 35 supersonic aircraft – "once Overture meets United's demanding safety, operating and sustainability requirements". Which is a big "if" – as *Overture* has not even had its design freeze at the time of the order, which had United pay an undisclosed sum up front. If things are working out according to plan, United wants to fly the first paying customers in its supersonic airliner configured with 88-seats from about 2029, mostly on North Atlantic routes from its Newark hub to Europe, such as to London (in three and a half hours) and Frankfurt (four hours). "A move that facilitates a leap forward in returning supersonic speeds to aviation", touted United. In any case, it was a very welcome shot in the arm for Boom, so any doubters seeing dark clouds appear over the return of supersonic airliners after Aerion's exit would proven otherwise. "*Overture* is the simplest, most doable and proven version of supersonic technology of which we think will find a big market," says Boom's CEO Blake Scholl. At the same time it is "one of the most complex, safety-critical machines ever built".

The planned list price is US$200 million, about 60% more than for a Boeing 787-9. Boom creates a "cost magic" out of this and speaks of a "speed dividend": as flight times are halved, goes the argument, the number of possible annual *Overture* flights are more than double that of a 787. With the supersonic airliner, up to 1327 long-haul trips become possible in a year; a Dreamliner can only do 584. "You can indeed achieve cost advantages if you fly faster," claims Scholl. *Overture* was supposed to reach the highest speed of any civil airliner ever built, and for Scholl that's at the core of his concept. "I think the lower speeds aimed for by other projects are actually the problem," he told the author in 2017. "If you look at an airline schedule about what trips passengers can take, Mach 1.5 isn't good enough in most cases. You have to go at least Mach 2." This is the only way to do a day trip from New York to London for an afternoon meeting. "50 years after Concorde, we need to be faster, not slower. There is a huge need to fly this fast, the other proposed lower-speed concepts will run into some real challenges when you look what you can actually do with them," Scholl insisted. No longer – in the context of the United order in June 2021, Boom silently

The test aircraft Boom XB-1 in the computer-generated air flow simulation (Boom Supersonic)

First draft of the Boom supersonic airliner (Boom Supersonic)

acknowledged that *Overture* would reach just Mach 1.7. But the perceived advantage of even that speed could also work out in a negative way. As it is questionable whether there is enough demand to utilise the high potential *Overture* offers, and in a profitable way. Also, the high productivity of each individual aircraft means that airlines might buy less of them overall. But that doesn't deter Scholl. "We think we are going to make more *Overtures* than

Boeing has made 787s", he claimed in October 2020, when Boeing had delivered nearly a thousand *Dreamliners*.

The proof of how seriously this company, employing about 140 people in 2020, means business has already been presented in October 2020. In a virtual ceremony, Boom Supersonic revealed the world's first privately built and funded supersonic aircraft in its hangar at Centennial Airport in Denver, it had reportedly cost US$80

million to build the jet called XB-1. In April 2020, the fuselage was matched with the wings, the last and most decisive step of assembly, for which heat and pressure is applied to the composite materials used. The rule of thumb is that it takes one year from XB-1's reveal to first flight, and when this book went to print the plan was still to have the first flight during 2021. Interestingly, in its evolution, the demonstrator officially nicknamed *Baby Boom* morphed

from a two-seater into a single-seat aircraft, which the company never publicly commented on.

Even the look-through renderings on their website still showed a second pilot compartment behind the first. "Through developing XB-1, we have learned a great deal about how to optimize the aircraft and came to the conclusion that only one pilot is needed to fly XB-1 and collect flight data. Our other test pilot will be flying a chase aircraft to ensure the safety of

XB-1 and the other test pilot, as well as to observe the test flights," the company explained in early 2021 when asked. *Baby Boom* is only a third of the size of the planned airliner, whose concept and design is supposed to be validated by test-flying the smaller iteration from Mojave Air and Space Port in California above the Mojave desert. The XB-1 will be shadowed by a T-38 Talon trainer, that is also used for pilot training. "We'll be learning all kinds of things along the

way," hoped Scholl. "We'll test our materials and our propulsion flow path. We'll learn more about adapting a subsonic engine for supersonic flight. We'll test new technologies like our pilot's virtual sight line through the nose of the plane. And along the way we'll continue building our company and our culture," the CEO pointed out. The noses of both Boom aircraft will be heated up to 174°C in flight – about 50°C degrees more than Concorde's nose which was heated to 127°C during cruise, which was slower than Boom's designs. The composite material now used is sufficiently heat-resistant. For *Overture's* flight test programme Boom reportedly expects a two-year period of flying with six aircraft.

Programme timeline delays are a given, especially in supersonic projects, and Boom experienced the same with the XB-1. Currently, Virgin Atlantic Airways holds pre-orders for ten *Overtures*, Sir Richard Branson, Virgin's founder, had supported Boom from the start, also offering to make use of synergies with his space venture Virgin Galactic. Together with Japan Airlines's commitments and the 15 aircraft ordered by United, the total value of 45 *Overtures* is about US$9 billion. In 2017

In contrast to Concorde and Tu-144 with four engines, Overture will be tri-jet (Boom Supersonic)

Scholl had talked about 76 aircraft having been pre-ordered by five airlines, apparently the amount of pre-orders is volatile, no difference to Concorde here. It is unclear how much any possible long-term fall-out of the pandemic, such as less business travel, might affect the airline's supersonic market, which has not been properly defined yet in any case. Concorde at least suffered greatly from bankers, movers and shakers not flying after 9/11. Still, investors seemed to maintain their belief in Boom's business plan, as only in December 2020, the company gained unicorn status by raising another US$50 million in funding on the capital market, bringing the total valuation of the company to US$1 billion.

"We are now at the crossroads of what is technologically achievable and for what there is a commercial market," said Scholl. "Blake's crystal ball is that subsonic will become low cost carrier market and supersonic will become the premium." Here again, we see unwavering optimism, or you could say wishful thinking, in terms of market projections. Boom assumes a huge surge in demand due to the new opportunity to fly supersonic for fares more in line with current subsonic premium fares. "Initially there will be much more demand than supply, meaning prices will be high at first," he predicts.

Up to 75 passengers can fly on Overture with just one seat on each side of the aisle, bit lie-flat seats (Boom Supersonic)

"*Overture* is an airplane that is affordable to tens of millions of people who fly business class today. But there is going to be *Overture 2* and *Overture 3*. We will make the aircraft progressively bigger, faster and more affordable. My aspiration is to make the fastest flight also the most affordable and ultimately get to a place where you can go anywhere in the world in four hours for a hundred bucks", states Scholl. Reportedly, once production has started, up to 60 *Overtures* a year could be produced. Scholl now sees a market for 1000 to 2000 aircraft, earlier he predicted 1300 units within ten years and founded this on studies by neutral analysts. Such statements are never reliable, most experts deem them to be totally unrealistic. "History shows that when flights get faster, people travel more often. It happened when jets replaced propeller aircraft, then traffic grew like six-fold. You will see a big increase in demand," Scholl told the author.

Boom Supersonic at least

Overture will not feature a droop nose. Live feed from a forward camera will enable the pilot to look down in front of them (Boom Supersonic)

chose to venture into the unknown as commercial enterprises have never dared to before, and the first privately built supersonic aircraft is only supposed to be the beginning. The probability that the demonstrator, already existing, will actually take off, can considered to be very high, all else remains speculation. Blake Scholl will celebrate his 40th birthday in 2021, which could coincide with being the year of *Baby Boom's* first flight. His motto is: "We are building a faster future faster." Since childhood he has been fond of aviation and has held a private pilots license since 2008, but he has never flown on Concorde to his big regret. "I was 23, working at Amazon in Seattle when it was shut down, and I didn't have that kind of money," recalls the current Boom CEO. Until 2014, he was active in other parts of the IT and internet industry, before he sold his company for e-commerce apps to the giant Groupon and thereby gained the launch capital. "I'm an internet guy by training, and airplane guy by passion," says Scholl. "People underestimate what you can do if you're committed and passionate."

Back in 2007 it struck him that there were no serious efforts being made anywhere to build a new supersonic airliner. "I thought: Could I start a company to build a supersonic jet? I thought the answer would be 'No'," remembers Scholl. "The more I learned the more I realized that we actually can do it. It's hard but it's possible, as in fact we already have all the key technology we need." The company Boom Supersonic, co-founded by Scholl in 2014, doesn't intend to re-invent the wheel, deliberately not, so as to keep its project realistic. Boom intends to take up proven technologies and put them together with today's tools to a new modern overall concept, 50 years after Concorde. "We've got better ways of optimized aerodynamics, we have new materials like carbon-fibre composites, we have significantly quieter and more efficient engines. If you take all these things together you can build a new-generation supersonic airplane that costs 75% less to fly than Concorde,"

Scholl sums it up. "And once we realized that we said: Either we have no courage, or we wanna start this company and do it."

Compared to the development period of Concorde in the 1960s there have been quantum leaps in efficiency made if anyone wanted to do the same all over again. In the 1960s, aerodynamic variations had to be tested in the wind tunnel, it could take half a year to change a few details in a design. "Today's computer simulations, in contrast, are really good, you can do all your optimization in the computer, meaning you can do six months of work in half an hour. And you can test thousands of more design ideas than you could before, which means you can find better ones and you end up with a refined aerodynamic

Originally, as in this early draft, the test aircraft XB-1 was configured as with two seats (Boom Supersonic)

Boom CEO Blake Scholl presents an Overture model in the livery of his airline partner Japan Airlines (AS)

shape of the airplane, with more lift and less drag," explains Scholl. These gains in efficiency have been reflected already in Boom's first design renderings. "*Overture's* fuselage is a little bit thicker at the front and thinner at the back, it's a very complex shape. If you try to make that out of aluminium, of which aircraft were all made until relatively recently, you can't do it. But with composites, you mould the parts into any shape you want them," beams Scholl. So far, the question of propulsion hasn't been solved for *Overture*; Rolls-Royce has been announced as a partner, which doesn't necessarily mean that's where the engines will ultimately come from.

That was easier on the XB-1, which is fitted with three engines of the military type General Electric GE J85-15 without afterburners, whose base model is from 1960 and built into the Northrop F5 fighter among others. "*Baby Boom* will use a small old engine, because that's what fits the airplane," Scholl admits. The design of the tri-jet had initially been revealed in 2016 as a two-seater, then first flight was planned for the end of 2017. But there were constant changes in the design and the engines. Until November 2018 the XB-1 was tested in the wind tunnel in a total of three campaigns, one of them being dedicated to the supersonic configuration of the engine inlets. This had still taken a full decade for Concorde.

Original wind tunnel model of the test aircraft XB-1 (AS)

on long over-water routes. But apparently, even the final number of engines is not a given yet. Current renderings still show three engines, but the company insisted in spring 2021: "We have not made a decision on the jet configuration yet." The turbofan engines were to be adaptions of a currently existing type with medium bypass ratio, otherwise being found on wide-body aircraft. They were supposed to enable a range of 8300 km, but in the context of the United order, Boom announced that range would now be only 4250 nautical miles (7871 km). In 2019 Scholl admitted that besides existing propulsion, Boom might also look at new developments. "Not a new technology, but a

Many hours were spent in the wind tunnel before the final configuration of the XB-1 was frozen (AS)

That's the reason Blake Scholl points out that, in this critical feature, the Boom aircraft will surpass Concorde by 5%. *Baby Boom* is 21 metres long, its wings span 5.20 metres and it has a maximum takeoff weight of 6100 kg. It is initially tested for Mach 1.3, "we are exploring additional speeds for XB-1s flight test programme", Boom announced in summer 2020. Before, it was saying that Baby Boom was supposed to reach Mach 2.2, exactly as *Overture* was later expected to, and to have a range of 1900 km. With heat exchangers the friction heat of the fuselage will be dissipated into the fuel.

Planning for *Overture* began in March 2016, with the initial configuration of just 30 seats soon being extended to hold 40 seats, then stretched to a length of 46 metres carrying 50 passengers and in 2019 being stretched again to the final 51.80 metres, now fitting in up to 75 travellers. The wingspan only grew marginally to 18.30 metres. Initially the airliner was being planned with just two engines, until in October 2016 a third one was added with inlets on both sides of the aft end of the fuselage. This was to not burden the aircraft with restrictions for two-engine jets

Test pilot Chris Guarente is preparing actual flights on the XB-1 from 2021 earlier in the simulator (Boom Supersonic)

on a case-by-case base, also according to the outcome of an environmental impact evaluation.

Especially in recent times and a context of flight shaming debates, it has become overly clear that future supersonic jets would have to fulfil very high environmental standards. And even if they do, this still would not guarantee the broad public's acceptance of such a new generation of aeroplanes. "We have realized that we need to make our aircraft environmentally compatible," reacted Blake Scholl. "Initially we said we would reach the efficiency of subsonic business class. But now we have realized that there will be an explosion of supersonic traffic, that's why we in fact have to be better than that." The question is if that really will appease environmental protectionists and climate activists. Boom is already putting concrete sustainability measures in place. Engine tests for the XB-1 have been done with sustainable aviation fuels as the whole flight test programme will be, also carbon emissions will be offset. *Overture* is conceived in a way that it will fully run on sustainable aviation fuels, a first in a new aircraft

new design would be possible." Existing turbofans in this class are basically too big for the smallish Boom aircraft, but are supposed to deliver the extra boost needed for supersonic flying.

For this, the subsonic engines would have to be modified and the rotating fans sucking in the air shrunk considerably. "We need a compromise between low airport noise and efficient fuel burn at cruise speed," explains Scholl and promises another quantum leap compared to Concorde: "The engine noise at the airport will be 30 times quieter than Concorde, which was really loud. And we don't use afterburners on takeoff, which is a big factor. And on the sonic boom side, it is also quieter by a factor of 30 versus Concorde, so the Boom passenger airliner will be 30 times quieter both at the airport and

with the sonic boom it causes," Boom's founder announced.

The current business model of Boom is still based on the assumption of flying supersonic just over water. "We shouldn't scratch our heads whether we can also fly supersonic over land but focus on over-water routes," says Scholl, but Boom was among the companies urging the US DOT in 2019 to lift the current ban. The Denver-based manufacturer was mostly focusing on being allowed test flights over land, Boom argued the ban was economically not comprehensible and undermined safety. Especially with the company located far away from the ocean it feared unreasonable costs if tests could only be conducted over water. In spring 2021, the FAA issued a new rule for manufacturers to apply for tests over land, granted

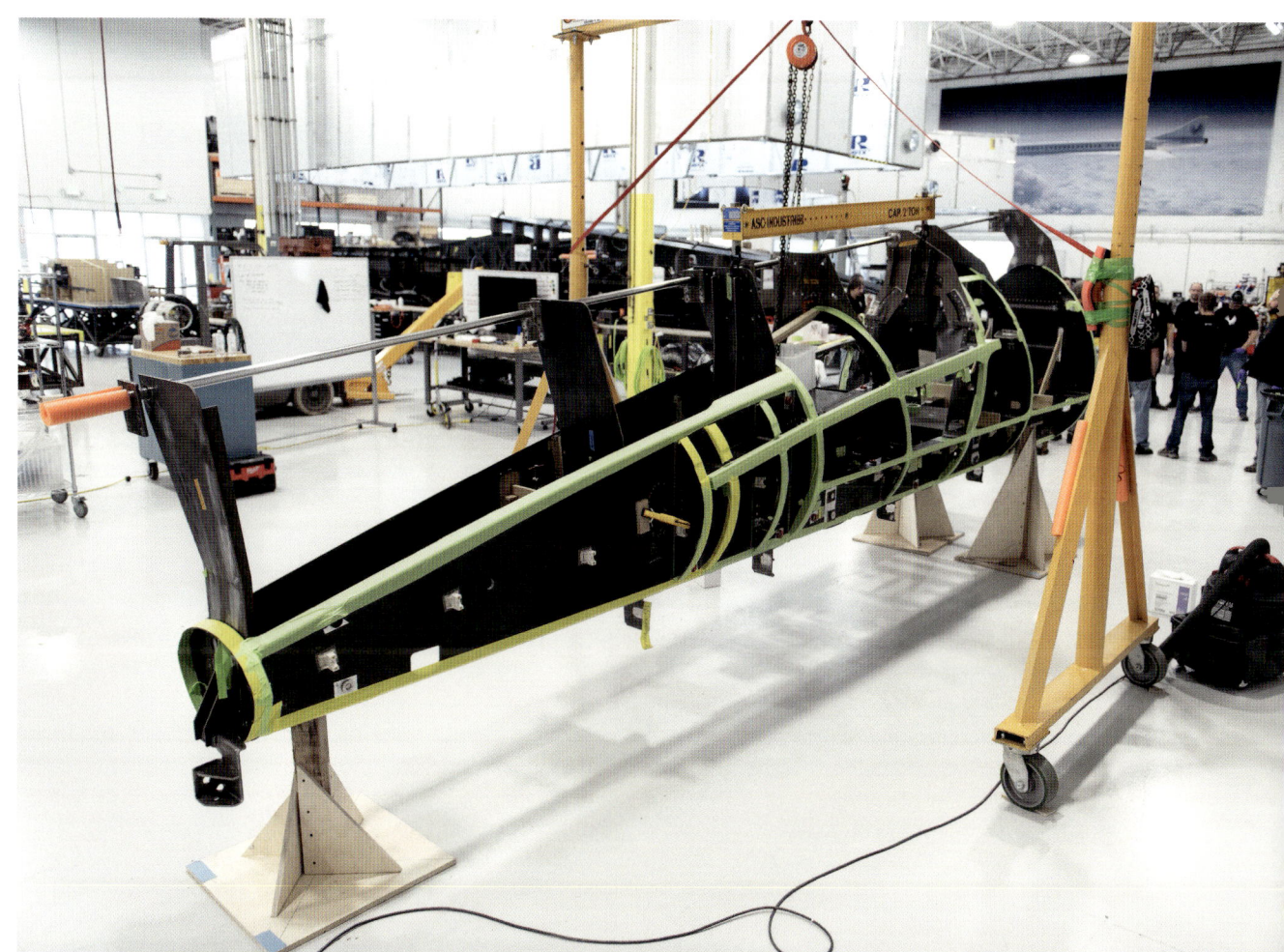

The carbon-fibre cockpit section of the XB-1 is glued together in the Denver hangar (Boom Supersonic)

Carbon-fibre cockpit and fuselage structures of the XB-1 were glued together in Denver and then "baked" in the autoclave (Boom Supersonic)

will be just one seat on each side of the aisle per row. Boom promises a "dramatically better on-board product" – including two lavatories. Obviously the early airline partners Virgin and JAL have convinced Boom that the precious cabin floor has to generate more revenue and more space is needed. Blake Scholl's mantra is: "The luxury is not the airplane, the luxury is on the ground and what you otherwise wouldn't get to do," as he told the author. He is rightly assuming that on a three-and-a-half hour transatlantic flight a one-class cabin is sufficient, whose comfort levels will be on par with a regional first or business class

programme. CO_2 emissions per passenger, promises Scholl, will not be higher than on today's long haul flights. Still, even that might not be enough to win over critics. While United touted *Overture* as "the world's first net-zero carbon supersonic aircraft", not only environmentalists, but serious industry observers call these "green-washing marketing claims." Pointing out that a four-hour supersonic transatlantic flight would still emit three to five times as much CO_2 as a subsonic flight on the same route. Even the respected Ascend consultancy summed it up very sceptically: "There are huge technical, financial, environmental and certification challenges for Boom, and it is an arguably strange decision from an airline that has recently made much of its environmental and sustainability targets." The timeline appears questionable to the analysts as well: "It seems extremely doubtful a new SST will enter service this decade."

The passenger experience on board *Overture*, operating with two pilots and up to four flight attendants, is supposed to be luxurious compared to Concorde, while space is still limited. Concorde's cabin equals today's premium economy class in its seat comfort. Using cutlery brought your elbow as close to that of your seatmate as is the case on a holiday airline. Not so on *Overture* – as there

The XB-1 wing elements made of light and strong composite material are assembled in Denver (Boom Supersonic)

Cutaway drawing of the XB-1 test aircraft (Boom Supersonic)

in today's wide-body aircraft.

Meaning there won't be lie-flat seats, due to space restrictions and mostly because supersonic flights are comparatively short. For night flights over the Pacific including a fuel stop, Scholl can imagine a separate first class cabin, twice as big as business class, with a seat pitch of 190cm and lie-flat seats, with two instead of one personal cabin stowage bins and two instead of one windows per seat. A cocoon with a focus on comfort and privacy, as Scholl puts it. One of the leading cabin designers, Anthony Harcup from Acumen Design in London, finds the passenger experience in a modern supersonic airliner is a challenge: "Designing a contemporary supersonic interior is a very different brief from that of contemporary commercial and private jet interiors. Rather than distracting the passenger from the idea of flying, the supersonic flight itself is the experience," said Harcup. "Passengers paying a top-end fare for a more limited footprint of living space will expect the same level of sophistication and product integration afforded to first and business class passengers on commercial flat-bed products."

Chances to have a separate first class cabin in such a relatively small aircraft might be even less than the probability that *Overture* will fly at all at some point. In any case, if the design is becoming reality, there will be spectacularly big windows, compared to

Conceptual render

United Airlines has committed to order 15 Overture, subject to the aircraft meeting the airline's specifications (Boom Supersonic)

Concorde's tiny portholes. The significantly stronger new composite material still provides enough structural stability, even without massively reinforced window frames. Eventually it might take until 2030 at least to finally put Boom's ambitious plans into reality and see *Overture* actually take off in passenger service. But Blake Scholl already imagines his first supersonic flight ever on it: "It's gonna be awesome, I can't wait to see the curvature of the earth for myself from 60,000 feet through our big windows. It's going to open the eyes of the world to a faster future." When it will be time for it, and if at all, is all but certain, despite Boom's dedication and that of its charismatic CEO. But it is certain that since Concorde, no other project has ever come as close to reaching the milestone of a launching a new generation of supersonic airliners. It also seems very likely that from today's status quo, still another decade might pass until commercial supersonic flights might be bookable again. Blake Scholl knows his quest is bold. "You have to rely on conservative assumptions in all fields to not overburden the business with risks," says Boom's founder. "Either we fail or we change the world."

Conceptual render

If the United order for supersonic aircraft is just a PR stunt or the start of a new era remains to be seen (Boom Supersonic)

gallery

"*The fascination with supersonic air travel is just a symbol of the human quest to go faster, higher, further.*"

Dr. Bernd Liebhardt, scientist focusing on high-speed air transport

Rollout of G-BSST on September 19th, 1968 (Collection Ad Jan Altevogt)

F-WTSS British Aerospace-Aerospatiale Concorde 001 F-WTSS at roll-out ceremony 11.12.1967(© Ad Jan Altevogt.)

F-WTSS British Aerospace-Aerospatiale Concorde 001 runway barrier trials.(©coll. Ad Jan Altevogt)

British Aerospace-Aerospatiale Concorde mock-up at Le Bourget 1967 (©Ad Jan Altevogt)

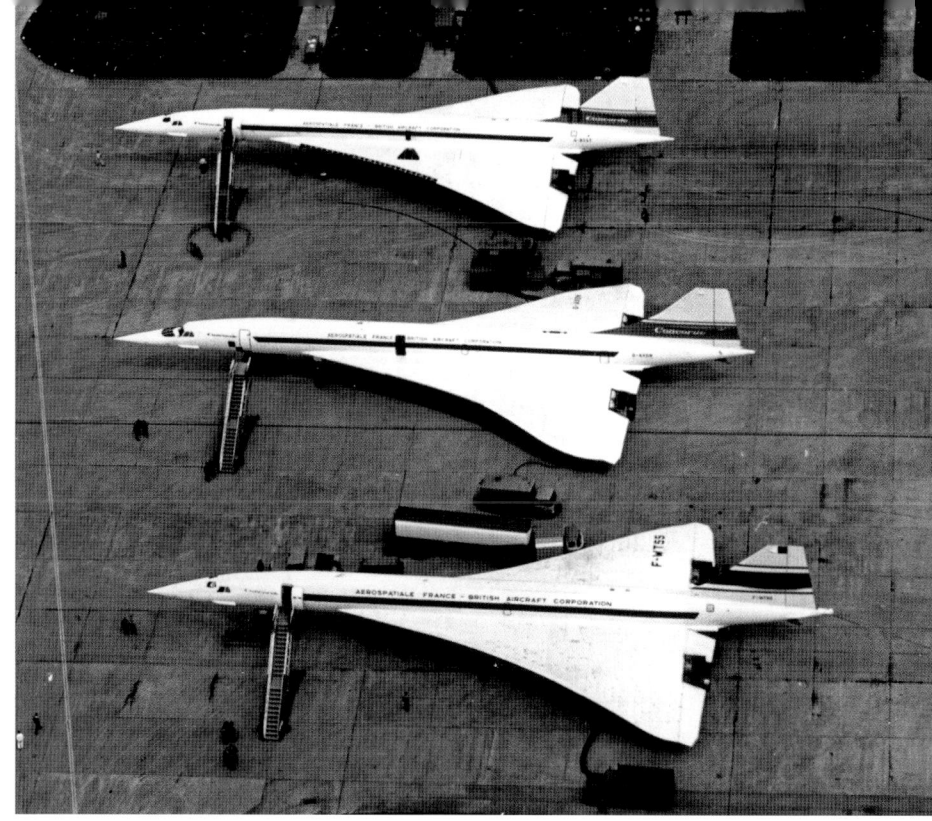

G-AXDN British Aerospace-Aerospatiale Concorde 001 F-WTSS and 01 G-AXDN and 002 G-BSST (©coll. Ad Jan Altevogt)

British Aerospace-Aerospatiale Concorde Mock-up Le Bourget June 1967 (©Ad Jan Altevogt)

During route proving flights in 1974, a test aircraft in British Airways livery flew to San Francisco for handling tests at the passenger terminal (Collection Ad Jan Altevogt)

A Concorde airframe is moved between hangars at final assembly in Toulouse (Rowland White)

Back in 1961 Lufthansa showed its very own concept for a supersonic airliner, aiming at carrying up to 140 passengers with Mach 3 (LH)

Concorde model in the old
Lufthansa livery in 1967 when
the German carrier secured
three options for the supersonic
airliner (LH)

Der Lufthanseat

Zeitung für die Mitarbeiter der Lufthansa   Nummer 98   13. Jahrgang   Februar/März 1967

Concorde für Lufthansa

Die Lufthansa hat sich Lieferpositionen für drei Überschallflugzeuge vom Typ Concorde gesichert. Diese in britisch-französischer Gemeinschaftsarbeit von British Aircraft Corporation und Sud Aviation gebauten Flugzeuge sollen der Lufthansa 1973, zwei Jahre nach dem Einsatz der ersten Serienflugzeuge bei anderen Luftverkehrsgesellschaften, geliefert werden. Die Lufthansa bleibt damit ihrem Grundsatz treu, nicht unter den Pionieren des Überschallverkehrs zu sein.

Entgegen früheren Erwartungen kann mit dem Einsatz amerikanischer Überschall-Verkehrsflugzeuge bei der Lufthansa nicht vor 1976 gerechnet werden, das heißt fünf Jahre nach Beginn der Concorde-Auslieferung an die ersten Kunden. Diese unerwartet lange Zeitspanne

hat zur jetzigen Entscheidung der Lufthansa beigetragen. Die bereits im März 1964 genommenen Optionen für drei US-Überschallflugzeuge bleiben jedoch bestehen. Der Stückpreis der Concorde beträgt 64 Millionen DM, der gesamte Anschaffungspreis der drei Flugzeuge einschließlich der Ersatzteile liegt bei rund 340 Millionen DM. Die Concorde wird ab 1973 Lufthansa-Passagiere in etwa vier Stunden von Frankfurt nach New York befördern können.

Die Lufthansa hat die Entwicklung dieses ersten zivilen Überschallflugzeugs der westlichen Welt periodisch neuen Studien unterzogen. Anfängliche Bedenken der Lufthansa wurden durch Konstruktionsänderungen, die die Hersteller inzwischen vorgenommen hatten, zerstreut. Ein Kaufvertrag wird erst abgeschlossen, wenn die erste Concorde ihre Flugtüchtigkeit bewiesen hat und die von der Lufthansa-Technik geforderten Leistungsbedingungen erfüllt werden. Die Concorde wird drei Jahre er-

probt werden, bevor die Hersteller bei den Luftfahrtbehörden die offizielle Zulassung für den zivilen Luftverkehr beantragen, die für Anfang 1971 erwartet wird.

Die Concorde wird mit einem maximalen Startgewicht von 158,8 Tonnen rund 8 Tonnen schwerer sein als eine Boeing 707. Sie wird 128 Passagiere in einer Reiseflughöhe von 16 500 m mit einer Geschwindigkeit von 2330 km/h befördern können.

Technische Daten	Concorde	Boeing 707-330B	Boeing SST
Spannweite (m)	25,60	44,42	54,96/32,23
Länge (m)	58,23	46,61	93,27
Höhe (m)	11,58	12,94	14,71
Maximales Startgewicht (kg)	158 760	150 414	302 090
Maximales Landegewicht (kg)	98 880	97 524	195 050
Sitzzahl	128	164	286
Kraftstoffmenge (l)	105 000	90 300	212 500
Reisegeschwindigkeit (km/h)	2 330	900	2 870
Maximale Reichweite (km)	6 671	9 060	7 282
Maximale Nutzlast (kg)	11 984	19 590	31 457
Startbahnlänge (m)	2 940	3 300	2 120
Landebahnlänge (m)	2 140	2 150	2 065
Reiseflughöhe (m)	16 500	11 000	19 200
Standschub der Triebwerke (kp)	4 x 15 920	4 x 8 165	4 x 28 668
Typ	(Bristol Siddely/ SNECMA)	(Pratt & Whitney)	(Gen. El.)
	Olympus 593	JT 3D-3B	GE 4/J5P

In March 1967 the Lufthansa staff paper announced the airline had signed three Concorde options (LH)

The Tu-144 prototype giving its Western debut at the Hanover Air Show in Germany 1971 (Collection Ad Jan Altevogt)

For the test campaign with NASA the Tu-144L received a new paint job in 1996 (AA)

CCCP-77112 displayed at Sinsheim now comes in for a landing using a parachute. It was the only airliner depending on this device as it had no reverse thrust (AA)

CCCP-77108 on display at the Samara Aviation Institute in 2002 (Ad Jan Altevogt)

This perspective makes it clear how important the extendable canard wings were for stable flight at low speeds (AA)

Tu-144 CCCP-77106 last flown in 1980 is now on display the Russian
Central Air Force Museum in Monino near Moscow (Andrey Khachtryan)

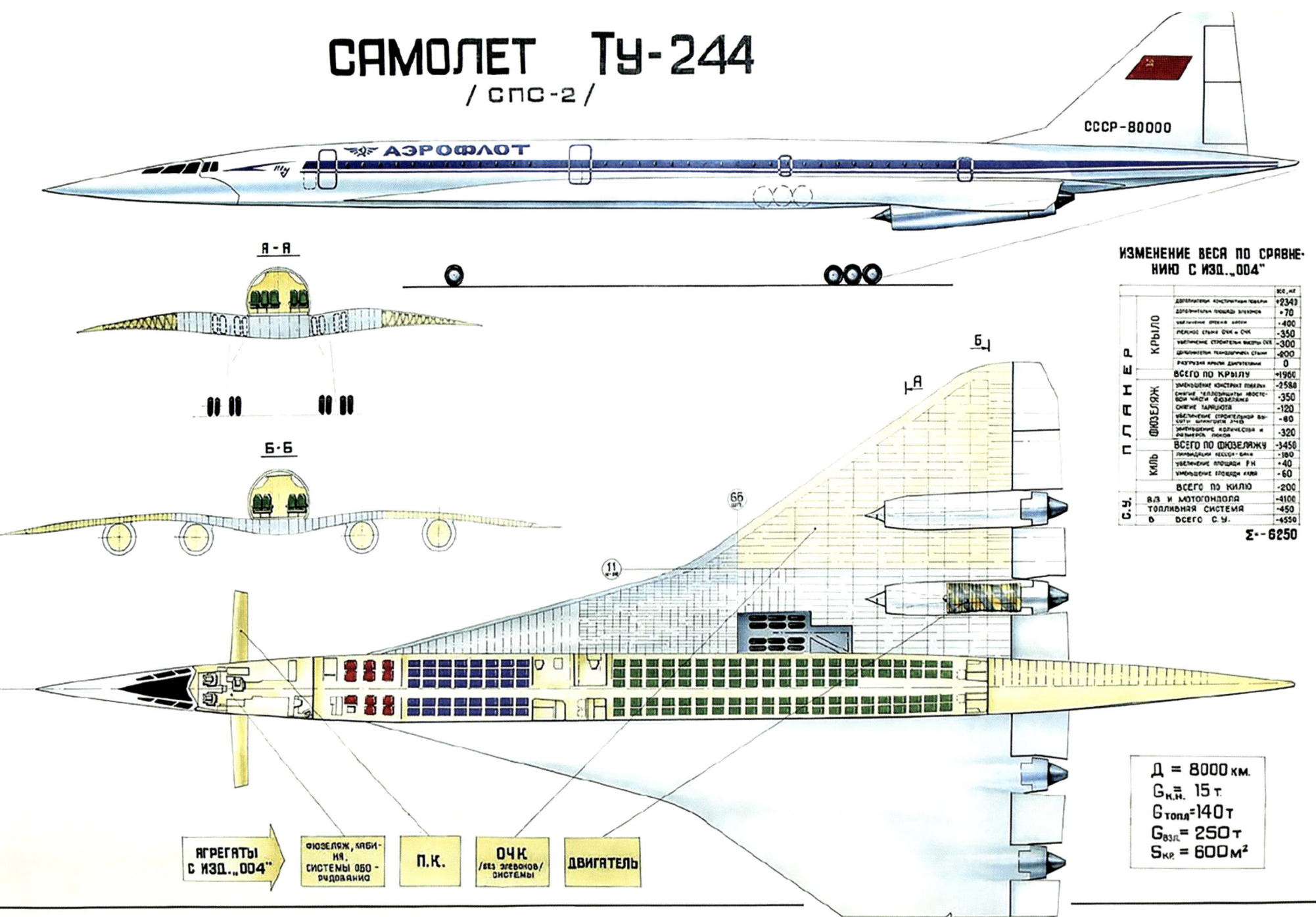

Draft for a successor model called the Tu-244 (Pavel Sineokiy)

For just two weeks in 1996, Concorde F-BTSD was painted in this promotional livery. The blue colour was not heat resistant and started to peel off during supersonic flight

The class of 1969: Concorde G-BOAC meeting a KLM Boeing 747-400 in Amsterdam in 1994 (Ad Jan Altevogt)

Approach checklist

Cabin crew call – 15 Min.
Landing briefing – Updated.
Taxi turn lights – On.
RAD/INS sws – RAD.
Flight deck door sw – Open.
Seat belt signs – On.
Engine rating mode – Take off.
Brake fans – On.
Engine recirculating valves – Shut.
Engine control schedule – Approach.
Engine feed pumps – All on.
Cross feed valves – Shut.
SSB – As required.
Batteries/d.c. split sw – As required.
Fuel/weight/CG – Checked.
ASI bugs – Update.
Seats & harness – Locked, pwr off & sec.
Autopilot changeover – Checked.
Visor/nose – 5 degrees.
Altimeters – QNH set/update.
10,000 feet – Landing lights on.
Nav 1 – Set ILS frequency.
AFCS – Set ILS course in AP1 and AP2.
9,000 feet – Alt hold on – set final App Alt/Alt AQC on.
Speed – 210 KIAS.
AFCS – When ILS active press VOR loc.
Visor/nose – down 12 ½ degrees.
At Loc intercept – Press land at AF.
Speed – 190 KIAS.
Approach checklist – Complete.

Landing checklist

At glide slope capture – Gear down – 4 Green.
Speed – 170 KIAS.
AFCS – When land 2 light is on, AP2 for Auto-land/Land 3 light on.
Speed – 160 KIAS.

A collection of UK newspaper headlines a day after Concorde flights ended in October 2003 (Richard Walker)   *Concorde Landing checklist*

Three Concorde temporarily grounded in London-Heathrow after their commercial flights in October 2003 (Baz Glenister)

Author Andreas Spaeth on the BA terminal rooftop in November 2001 after resumption of Concorde services (AS)

Concorde test flight to Acapulco in 1974 (Anne Rebours)

Concorde test flight to Acapulco in 1974 (Anne Rebours)

Air France Concorde F-BVFA at the Ferte Alais air show in 1987 performing with the Patrouille de France aerobatic team (Ad Jan Altevogt)

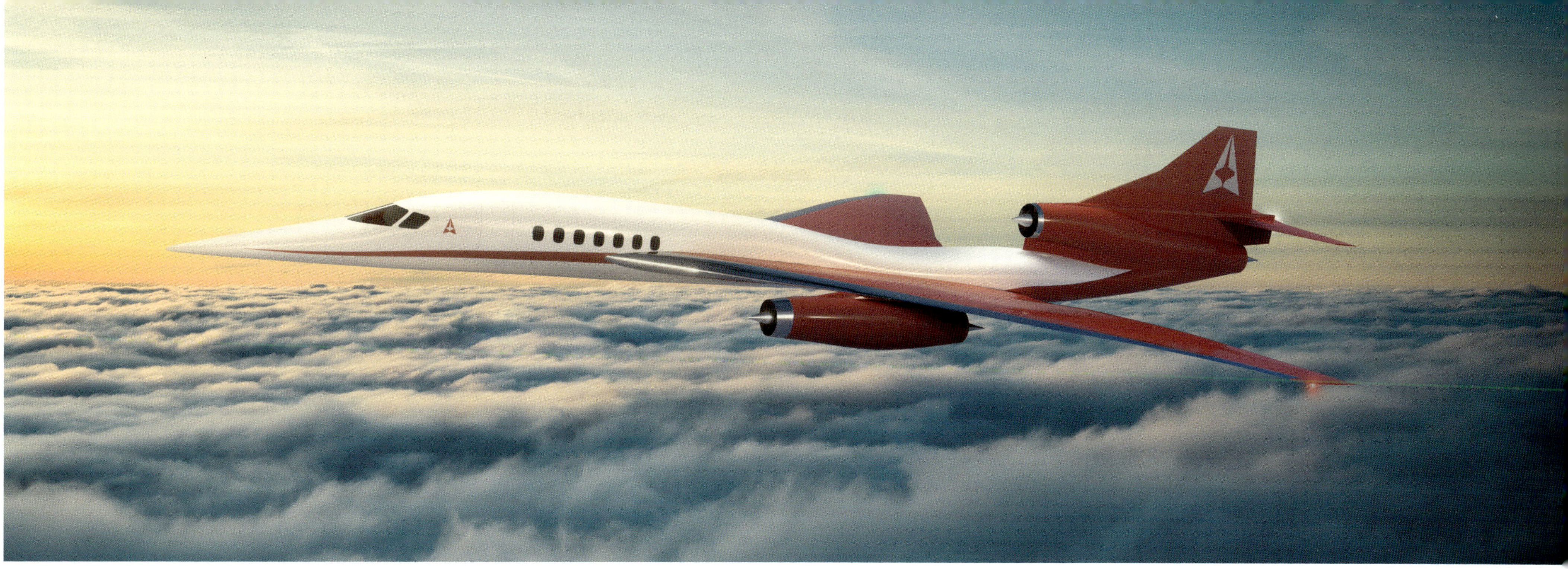

The last and final design for the Aerion AS2 supersonic business jet from 2020, before the company folded in July 2021 (Aerion)

The team at Lockheed Martin Skunk Works in Palmdale, California, merged the major sections of the X-59 Quiet SuperSonic Technology aircraft. This marks the first time the X-59 resembles an actual aircraft [Lockheed Martin via NASA]

Rendering of the X-59 QueSST on takeoff, displaying its unusual shape aimed at minimizing noise (NASA)

appendix

Virgin Galactic announced it was doing a feasibility study for a sustainable Mach 3 airliner with Rolls-Royce in 2020 (Virgin Galactic)

"Supersonic is all about prestige "- Interview with a scientist

Airbus then holding company EADS introduced a draft of a Zero Emission High Speed Transport (ZEHST) already in 2011 (AS)

An interview with Dr. Bernd Liebhardt who works at the institute for air transportation systems of the German Aerospace Centre DLR in Hamburg. As one of a few scientists, Bernd Liebhardt is focusing mainly on high-speed air transport and did his doctorate about this topic in 2016.

Why have people always been fascinated by supersonic air transport?

Liebhardt: It is just a symbol of the human quest to go faster, higher, further. Since the 1950s people have been stuck at Mach 0.8 on jet flights. That is difficult to accept in the human quest for progress, especially as it had been proven in the late 1960s that it is possible to travel at supersonic speed. So the fascination about supersonic flying is just human.

Dr. Ing. Bernd Liebhardt works about civil supersonic aircraft projects at the German Aerospace Centre DLR in Hamburg

How do you judge the significance of Concorde and Tu-144 for aviation history in hindsight?

These were both tremendous technological achievements. Concorde, you might say, was Europe's Apollo program. Europeans and Soviets have built the only supersonic airliners that ever flew. The fact Concorde was operating scheduled services accident-free from 1976 to 2000. Twenty-four years was an extraordinary achievement. It was a pioneer project, which unfortunately was much too large in scale as there was a belief in the unstoppable technological progress at the time, but it was fitting for its period. One that was probably a bit overambitious in what the aircraft was supposed to deliver, a bit smaller and slower would probably have worked out better.

In any case it has paved the way for supersonic research that has never stopped since.

What are the biggest hurdles on the way to a modern supersonic airliner?

The biggest challenges are the environmental problems caused by such aircraft, mostly their high fuel consumption and corresponding CO2 emissions at high altitudes. At this level the air is very dry and every jet engine produces water vapours that don't belong there. And of course the sonic boom continuously emitted by such aircraft wander to the ground, audible in a wide corridor underneath the flight path. That's why supersonic flights have been banned overland so far.

Currently there are big efforts under way to lift this ban in the medium term future, thanks to new technology. Is that a realistic option?

I only see that coming in the long term. Soon there will even be test flights conducted over cities. But there are other problems that cannot be solved that way, and as far as I know these are not solved yet. Once an aircraft accelerates to supersonic speed and then further, inevitably a super boom is created, quite an intense one, that hits the ground in a certain position and can cause damage. You cannot do anything about it, that happens every time when the aircraft accelerates to maximum speed. That is not changed by any new low-boom aircraft concepts which are focusing mostly on being efficient during cruise. Also unsolved is the question how the boom will be perceived inside of buildings.

And new aircraft designs can't change that?

No, any new low-boom shaped aircraft like the X-59 QueSST does not help initially, as the lower frequencies of the boom are decisive, and that you can hardly remove. Depending on atmospheric conditions at a certain place at a particular point in time, the boom can also sound much differently to what was envisioned before. There is a whole lot of technical problems to be solved. If maybe 90% of the people were disturbed by the boom emanating from Concorde, the new technology can maybe achieve that only 30-50% will be in the future.

One obstacle remains the acceptance by the population. Flights overland causing a boom will be difficult to push through even going forward, as supersonic flights do not

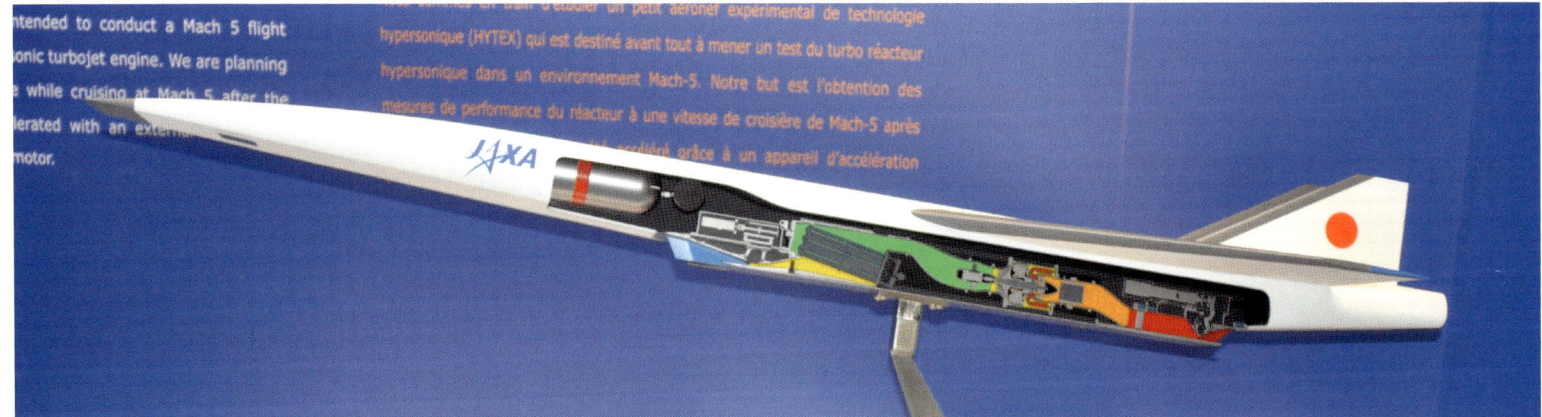

Japanese draft for a hypersonic HYTEX experimental aircraft (AS)

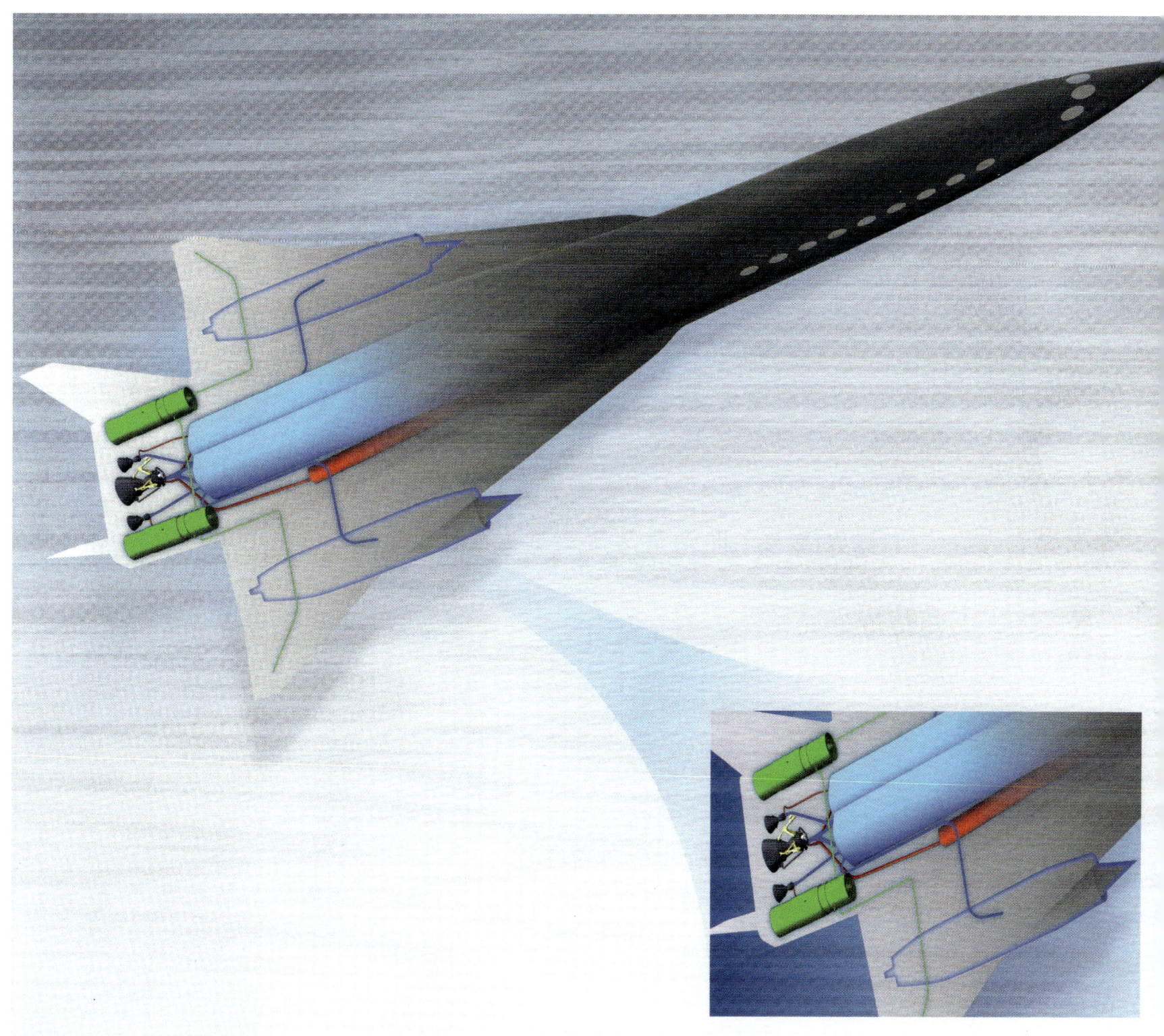

bring benefits to everybody and therefore there is no incentive to accept such a disturbance.

Lockheed Martin, building the X-59 QueSST, has also presented a forty seater concept based on the test bed. Would that be a promising path?

That is for the first time a reasonable size for a future supersonic airliner. It could be even smaller than these forty seats planned for now. As this aircraft could possibly fly supersonic overland its market is considerably bigger, only because of its more suitable size.

Otherwise there are not even enough city pairs for a 40-seater on which one could find enough passengers willing to pay for supersonic flights. A small aircraft is helpful both to fill it sufficiently, but also to lower the noise it emits, as the intensity of the boom depends on the aircraft mass.

What would be the ultimate size of a future supersonic airliner for you?

Virgin Galactic announced it was doing a feasibility study for a sustainable Mach 3 airliner with Rolls-Royce in 2020 (Virgin Galactic)

Somewhere between nineteen and thirty seats, as you need an additional emergency exit above twenty seats.

Boom Supersonic, by contrast, has repeatedly increased the size of its *Overture* design, do they have other insights?

I am under the impression that they have somehow been carried away a bit by their own hype. When I saw the Boom concept for the first time in 2015, it still had thirty seats. That was a bit far-fetched for me, and the intended cruise speed of Mach 2.2 was very ambitious. That would have meant the aircraft would have a much higher productivity than other aircraft. Once you have a certain market size and you serve it much faster, you need less aircraft, which in turn limits the amount of aircraft you can sell. If you hope to get lower seat costs by enlarging the aircraft to create a bigger market, then you have: a) a bigger aircraft that costs you more per flight; b) less revenue, as you have to lower ticket prices to enhance load factors; and c), hardly any passengers, as the market is not very sensible at that pricing. Overall, you lose money.

Blake Scholl, the founder of Boom, still speaks of a thousand or two thousand aircraft he wants to sell. Is that delusion?

Boom Supersonic is possibly not selling an aircraft, but a hype. According to our analyses, the market for their aircraft is probably much smaller than they claim. Boom says they will create a new passenger market, but this ignores the fact that markets for luxury goods such as supersonic flights are relatively static. Low cost carriers are developing new markets, offering a commodity product more cheaply than before, but this price elasticity doesn't apply here. No matter if they adjust the price for a supersonic ticket slightly, raising it, lowering it or keeping it static – the demand will not change a lot. They might be able to generate a bit of extra demand, but definitely not enough to fill the aircraft.

But that means that there actually isn't any lucrative market for supersonic passengers?

Yes there is, we see a very tiny window for it. For airliners with nineteen seats, where the manufacturer can sell the same airframe and cabin size also as a business jet. This is a thesis we have promoted for quite some time. It would mean putting a foot in the market with an aircraft you do not have to follow up with a second expensive development. To serve not just major city pairs, but also secondary markets economically, simply because you need so few passengers to fill the aircraft. We have calculated the ticket cost in a nineteen-seat airliner – that

In the draft for the ZEHST aircraft most of the cabin volume is taken up by hydrogen tanks in blue (AS)

would be in line with what you have to pay today for Business or First Class tickets.

What drove countries then and what drives private companies today to try to realize such a project?

It's all about prestige, as it has been since the moon landing. Not about spin-off products or lucrative markets, but about prestige. That was the case with the moon landing, with Concorde and the Tupolev Tu-144. Even in today's state-funded supersonic research efforts this still plays a major role. Doing such research, it is often argued, is about gaining time, but main drivers are the human pioneering spirit and the gain in prestige for anyone getting something like this on the way. And all the unrealistic market projections are ways to justify working on this. There is a market but it is very limited, and it is unclear if it is big enough to have all participants gain in the end. You have to be willing to take risks to push through something like this.

What will be the next step after supersonic - hypersonic flights?

From my point of view there seems to be a fairly major focus currently on hypersonic research in the Western world, for flying above Mach 5.

Possibly because the military is interested to equalize the perceived advantage of Russia and China. Concerning passenger transport, for me hypersonic lies several decades into the future. Supersonic is already a very difficult field to master, and hypersonic goes another big step beyond even.

Initially major technological problems such as surface heating and propulsion stability have to be solved, only then it will be possible to assess at some point if this is even suitable as a transport platform for humans and cargo. First one should concentrate on realizing sustainable supersonic flight, then it's time to think about civil hypersonic operations.

Virgin Galactic was focusing on space tourism for a decade. If it can also realize a supersonic airliner is even less certain (Virgin Galactic)

Production lists and whereabouts of the Concorde

Number: 001	Registration: F-WTSS	First Flight: March 2, 1969	Last Flight: 19, October 1973	Flight Hours: 812
French prototype, on display at the Musée de l'Air, Paris Le Bourget, France			
Number: 002	Registration : G-BBST	First Flight: 9, April 1969	Last Flight: 4, March 1976	Flight Hours: 836
British prototype, on display at Fleet Air Arm Museum, Yeovilton, UK			
Number: 101	Registration : G-AXDN	First Flight: 17, December 1971	Last Flight: 20, August 1977	Flight Hours: 632
British pre-production aircraft, on display at Imperial War Museum, Duxford, UK			
Number: 102	Registration : F-WTSA	First Flight: 10, January 1973	Last Flight: 20, May 1976	Flight Hours: 656
French pre-production aircraft, on display at Musée Delta, Paris Orly Airport, France			
Number: 201	Registration : F-WTSB	First Flight: 6, December 1973	Last Flight: 19, April 1985	Flight Hours: 909
French production test aircraft, on display at Musée Aèroscopia, Toulouse Blagnac, France			
Number: 202	Registration : G-BBDG	First Flight: 13, December 1974	Last Flight: 24, December 1981	Flight Hours: 1.282
British production test aircraft, on display at Brooklands Museum, Weybridge, UK			
Number: 203	Registration : F-BTSC	First Flight: 31, January 1975	Last Flight: 25, July 2000	Flight Hours: 11.989
French production test aircraft, destroyed in accident at Paris CDG Airport July 25, 2000			
Number: 204	Registration : G-BOAC	First Flight: 27, February 1975	Last Flight: 31, October 2003	Flight Hours: 22.260
In service with British Airways, on display at Runway Visitor Park, Manchester Ringway Airport, UK			
Number: 205	Registration : F-BVFA	First Flight: 27, October 1975	Last Flight: 12, June 2003	Flight Hours: 17.824
In service with Air France, on display at Smithsonian Air and Space Museum Udvar Hazy Centre, Chantilly, Virginia, USA			
Number: 206	Registration : G-BOAA	First Flight: 5, November 1975	Last Flight: 12, August 2000	Flight Hours: 22.768
In service with British Airways, on display at National Museum of Flight in East Fortune, Scotland			
Number: 207	Registration : F-BVFB	First Flight: 6, March 1976	Last Flight: 24, June 2003	Flight Hours: 14.771
In service with Air France, on display at Technikmuseum Sinsheim, Germany			
Number: 208	Registration : G-BOAB	First Flight: 18, May 1976	Last Flight: 15, August 2000	Flight Hours: 22.296
In service with British Airways, parked at British Airways maintenance London Heathrow Airport, UK			
Number: 209	Registration : F-BVFC	First Flight: 9, July 1976	Last Flight: 27, June 2003	Flight Hours: 14.332
In service with Air France, on display in front of Musée Aéroscopia, Toulouse Blagnac, France			
Number: 210	Registration : G-BOAD	First Flight: 25, August 1976	Last Flight: 10, November 2003	Flight Hours: 23.397
In service with British Airways, on display at Intrepid Sea-Air-Space Museum, New York City, USA			
Number: 211	Registration : F-BVFD	First Flight: 10, February 1977	Last Flight: 27, May 1982	Flight Hours: 5.814
In service with Air France, scrapped in 1994			
Number: 212	Registration : G-BOAE	First Flight: 17, March 1977	Last Flight: 17, November 2000	Flight Hours: 23.376
In service with British Airways, on display at Grantley Adams International Airport, Barbados			
Number: 213	Registration : F-BTSD	First Flight: 26. June 1978	Last Flight: 14. June 2003	Flight Hours: 12.974
In service with Air France, on display at Musée de l'Air, Paris Le Bourget, France			
Number: 214	Registration : G-BOAG	First Flight: 21, April 1978	Last Flight: 5, November 2003	Flight Hours: 16.239
In service with British Airways, on display at Museum of Flight, Seattle, USA			
Number: 215	Registration : F-BVFF	First Flight: 26, December 1978	Last Flight: 11, June 2000	Flight Hours: 12.421
In scheduled service with Air France, on display at Paris CDG Airport, France			
Number: 216	Registration : G-BOAF	First Flight: 20, April 1979	Last Flight: 26, November 2003	Flight Hours: 18.257
In scheduled service with British Airways, on display at Aerospace Bristol Museum, Bristol, UK			

Technical data of the Concorde

Concorde	Prototypes	Pre-Production	Production
Wingspan (m)	25,56	25,56	25,56
Length (m)	56,20	59,13	62,13
Height (m)	11,10	11,58	12,22
Max. fuselage width (m)	2,88	2,88	2,88
Main gear track (m)	7,72	7,72	7,72
Wheelbase (m)	18,19	18,19	18,19
Cabin length (m)	39,57	39,57	39,57
Max. cabin width (m)	2,63	2,63	2,63
Max. cabin height (m)	1,96	1,96	1,96
Wing area (m²)	358,25	358,25	358,25
Elevon area (m²)	32	32	32
Rudder area (m²)	10,40	10,40	10,40
Passengers max	138	148	131
Engine	Olympus 593-3B	Olympus 593-4	Olympus 593 Mk610-14-28
Max. thrust w. reheat (kN)	153	156,1	169,3
Op. empty weight (kg)	61.510	68.950	78.900
Zero fuel weight (kg)	74.845	83.010	92.080
Max. takeoff weight (kg)	147.870	158.760	185.070
Max. landing weight (kg)	90.720	98.885	111.130
Max. payload (kg)	11.795	12.700	13.155
Fuel capacity (Liter)	119.786	---	---
Fuel burn in cruise (Liter/h)	25.700	---	---
Typ. takeoff speed. (km/h)	360	360	397
Typ. landing speed.(km/h)	300	300	300
Climb rate (m/sec)	25,4	25,4	25,4
Service ceiling (m)	19.800	18.300	18.300
Cruise speed.(km/h)	2.190 = M 2,02	2.190	2.190
Takeoff runway at 15 m	2.900	2.925	3.600
Landing runway at 15 m	2.300	2,400	2.200
Range with 9 t payload (km)	7.770	6.580	6.850
Range w. max. payload(km)	6.275	6.275	6.230

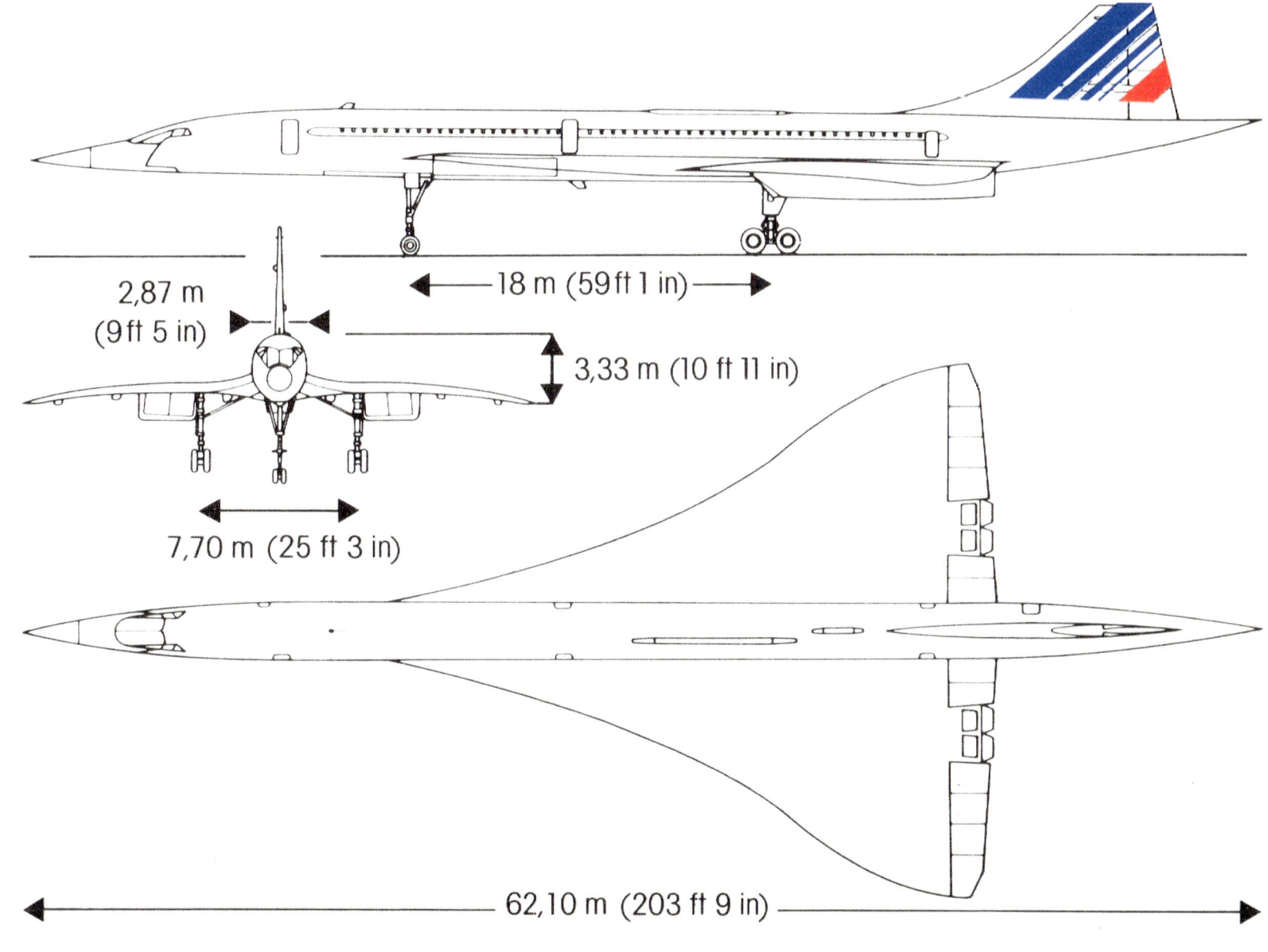

2,87 m
(9 ft 5 in)

18 m (59 ft 1 in)

3,33 m (10 ft 11 in)

7,70 m (25 ft 3 in)

62,10 m (203 ft 9 in)

Production lists and whereabouts of the Tupolev Tu-144

Number : 00-1	Registration: CCCP-68001	First Flight: 31, December 1968	Last Flight: 27, April 1973	Flight Hours: 180
Prototype, scrapped at Moscow Zkukovsky				
Number : 01-1	Registration: CCCP-77101	First Flight: 1, June 1971	Last Flight: unknown	Flight Hours: 339
First pre-production aircraft, scrapped				
Number : 01-2	Registration: CCCP-77102	First Flight: 20, March 1972	Last Flight: 3, June 1973	Flight Hours: n.b.
Test aircraft, crashed near Paris Le Bourget June 3, 1973				
Number : 02-1	Registration: CCCP-77103	First Flight: 13, December 1973	Last Flight: unknown	Flight Hours: 313
Test aircraft, scrapped 1984				
Nummer: 02-2	Registration: CCCP-77104	First Flight: 14. Juni 1974	Last Flight: unknown	Flight Hours: 431
Test aircraft, scrapped 1987 at Moscow Zhukovsky				
Nummer: 03-1	Registration: CCCP-77105	First Flight: 30, November 1974	Last Flight: unknown	Flight Hours: 314
Test aircraft, scrapped 1995 at Moscow Zhukovsky				
Number : 04-1	Registration: CCCP-77106	First Flight: 4, March 1975	Last Flight: 29, February 1980	Flight Hours: 582
First serial production aircraft, on display at Central Air Force Museum, Monino, Russia				
Number : 04-2	Registration: CCCP-77108	First Flight: 12, December 1975	Last Flight: 27. August 1987	Flight Hours: 68
Stored at Samara-Smyshliajevka Airport, Russia				
Number : 05-1	Registration: CCCP-77107	First Flight: 20, August 1975	Last Flight: 29, March 1985	Flight Hours: 357
On display at Kazan National Research Technical University, Russia				
Number : 05-2	Registration: CCCP-77109	First Flight: 29, April 1976	Last Flight: unknown	Flight Hours: n.b.
In service with Aeroflot, supposedly scrapped at Voronezh				
Nummer: 06-1	Registration: CCCP-77110	First Flight: 14, February 1977	Last Flight: 1, June 1984	Flight Hours: 314
In service with Aeroflot, on display at Museum of Civil Aviation in Ulyanovsk, Russia				
Number : 06-2	Registration: CCCP-77111	First Flight: 27, April 1978	Last Flight: 23, May 1978	Flight Hours: 9
Supposed to join Aeroflot, crashed May 23,1978 at Yegoryevsk on test flight				
Number : 07-1	Registration: CCCP-77112	First Flight: 19, February 1979	Last Flight: 12, November 1981	Flight Hours: 197
On display at Technikmuseum Sinsheim, Germany				
Number : 08-1	Registration: CCCP-77113	First Flight: 2, October 1979	Last Flight: unknown	Flight Hours: 223
Scrapped in 2001 at Moscow Zhukovsky				
Number : 08-2	Registration: CCCP-77114	First Flight: 13, April 1981	Last Flight: 14, April 1999	Flight Hours: 443
modified to Tu-144LL for Tupolev/NASA test programme, on display at roundabout close to main gate of Moscow Zhukovsky air base, Russia				
Number : 09-1	Registration: CCCP-77115	First Flight: 4, Oktober 1984	Last Flight: 12, May 1986	Flight Hours: 38
Stored at Tupolev in Moscow Zhukovsky				
Number : 09-2	Registration: CCCP-77116	First Flight: ---	Last Flight: ---	Flight Hours: 0
Never flown, remained as unfinished hulk when programme was cancelled in 1985. Supposedly scrapped at Voronezh				

Source: www.concordesst.com, www.144sst.com, British Airways

Technical data of the Tupolev Tu-144S

Tupolew Tu-144S	production version
Wing span (m)	28,80
Length (m)	65,70
Height (m)	12,85
Max. fuselage width (m)	3,80
Main gear track (m)	6,05
Wheelbase (m)	19,60
Cabin length (m)	26,50
Max. cabin width (m)	3,58
Max. cabin height (m)	1,93
Wing area (m²)	503
Passengers max.	140
Engines	4x Kuznetsow NK-144
Thrust with reheat (kN)	127 each
Max. takeoff weight (kg)	180.000
Max. landing weight (kg)	120.000
Max. payload (kg)	15.000
Operating weight empty (kg)	85.150
Cruise speed. (Mach)	1,88
Takeoff speed. (km/h)	355
Landing speed. (km/h)	280
Service ceiling (m)	20.000
Takeoff runway length (m)	3.000
Landing runway length (m)	2.600
Range with full payload (km)	3.080
Climb rate (m/sec)	25,4
Service ceiling (m)	18.300
Cruise speed. (km/h)	2.190
Takeoff runway at 15 m (m)	3.600
Landing runway at 15 m (m)	2.200
Range with 9 t payload. (km)	6.850
Range w. max. payload . (km)	6.230

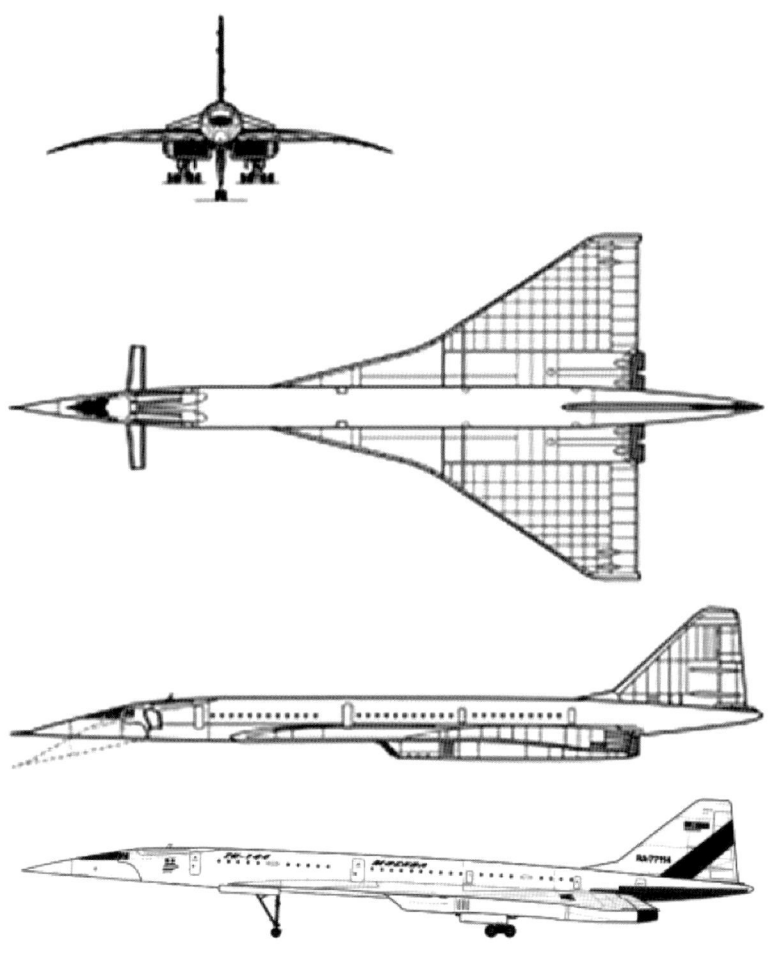

Technical data of the Boeing 2707-100 and 2707-300

	Boeing 2707-100 (1966)	Boeing 2707-300 (1969)
Wing span	32,23 m-54,97 m (variable sweep wing)	38,64 m
Length	90,83 m	96,01 m
Height	16,54 m	---
Max. fuselage width	---	3,45 m
Main gear track	8,53 m	---
Wheelbase	32,89 m	---
Wing area	856,6 m2	716 m2
Passengers max	277	296
Engines	4x GE4/J	4x GE4
Thrust with reheat	267 kN each	290 kN each
Max. takeoff weight	272.160 kg	306,175 kg
Max. payload	31.000 kg	31,000 kg
Weight empty	290.200 kg	130,400 kg
Fuel burn in cruise	14,200 litres/h without reheat	14,200 litres/h without reheat
	35,000 litres/h with reheat	35,000 litres/h with reheat
Cruise speed.	Mach 2,7	Mach 2,7
Takeoff speed.	293 km/h	---
Landing speed.	248 km/h	---
Service ceiling	18.300-21.300 m	---
Takeoff runway length	2.210 m	---
Landing runway length	1.950 m	---
Range w. full payload	6.440 km	---
List price	---	US$ 48 m (today US$ 337 m)

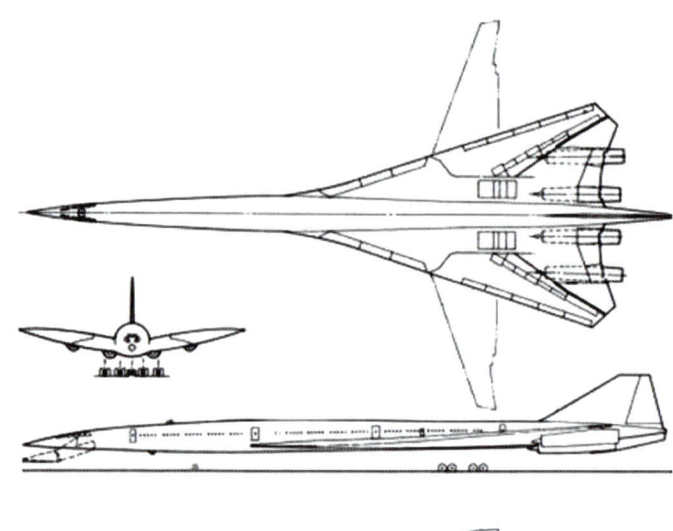

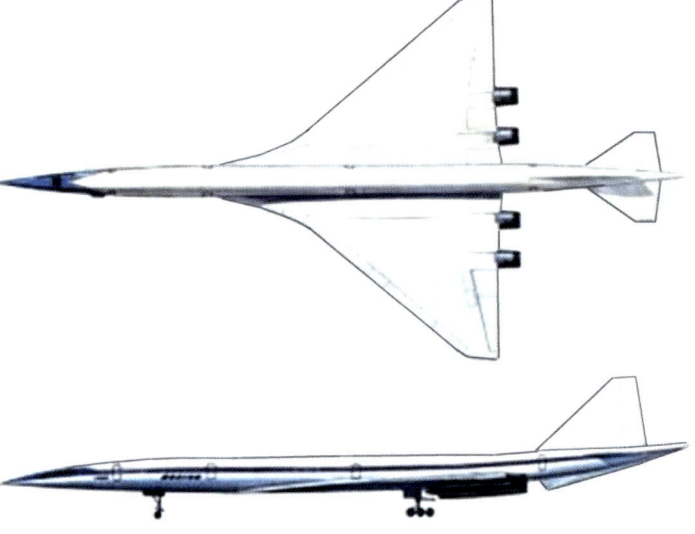

Lockhead Model 2000-7A specifications

	Lockheed Model 2000-7A (1966)
Wing span	35,36 m
Length	83,21 m
Height	14,27 m
Main gear track	7,36 m
Wheelbase	26,23 m
Wing area	875,5 m2
Passengers max	258-273
Engines	4x GE4/J5 or PWJTF17A
Thrust with reheat	je 272 kN
Max. takeoff weight	249.475 kg
Max. landing weight	154.220 kg
Max. payload	27.215 kg
Weight empty	107.900 kg
Cruise speed	Mach 2,7
Takeoff speed	295-325 km/h
Landing speed.	250-270 km/h
Service ceiling	23.317 m
Takeoff runway length	1.980m
Landing runway length	2.075m
Range with full payload	6.440 km

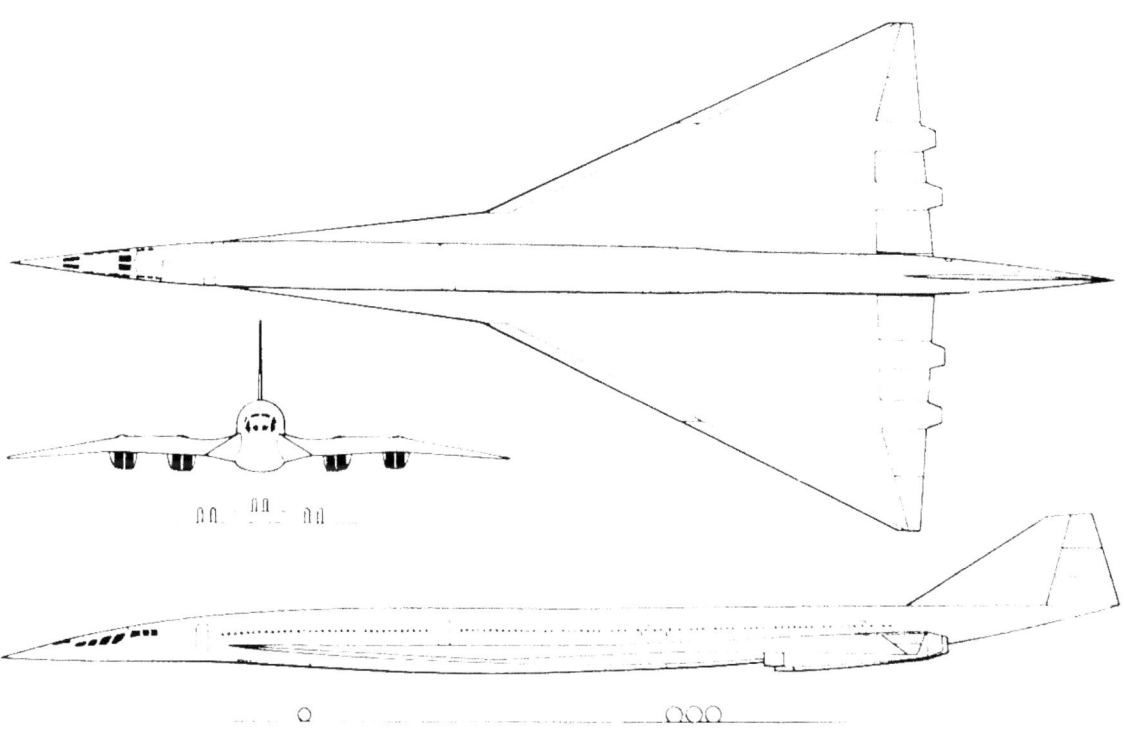

Technical data of the Aerion AS2

Aerion AS2	(Stand 2018)	(Stand 2020)
Wing span	23 m	24,07 m
Length	52m	44,17 m
Height	6,70m	---
Cabin length	9,10m	---
Max. cabin width	2,20m	---
Max. cabin height	1,90m	---
Wing area	140,4 qm	---
Passengers max.	8-12	---
Engines	3x GE Affinity Turbofan	---
Thrust	80 kN each	---
Max. takeoff weight	60.328 kg	63.049 kg
Max. payload	3.628 kg	---
Weight empty	26.762 kg	---
Cruise speed	Mach 1.4, subsonic Mach 0.95, boomless Mach 1.1-1.2	---
Range w. full payload	7,800 km at Mach 1.4, 10,000 km at Mach 0.95	---
Development cost	ca. 4 Mrd. US$	---
List price	120 Mio. US$	---
Market projection	300 Flugzeuge in zehn Jahren	---

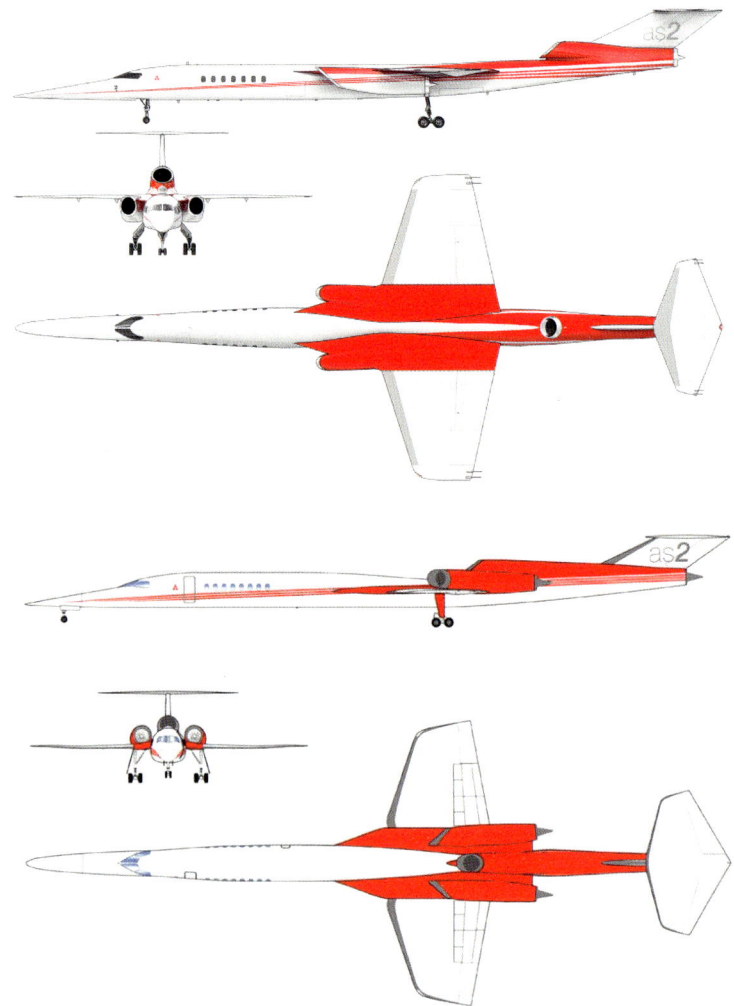

Technical specifications of the Boom Supersonic XB-1 and Overture

Boom Supersonic	XB-1	Overture
Wing span	5,20 m	18,3 m
Length	21 m	51,8 m
Passengers max.	2	55-75
Engines	3x GE J85-15 Turbojet	3x Turbofan
Thrust	19 kN each	67-89 kN each
Max. takeoff weight	6.123 kg	77.111 kg
Cruise speed	Mach 2,2	Mach 2,2
Nose temp. max.	153°C	153°C
Service ceiling	---	18.200 m
Takeoff runway length	---	3.048 m
Range w. full payload	1.900 km	8.300 km
Development cost	---	US$6 billion
List price	---	US$200 million
Market projection	Demonstrator	1000-2000 aircraft in ten years

Sources: www.concordesst.com, www.tu144.com, www.globalsecurity.org, www.wikipedia.org, Institute for Defence Analysis, Aviation Week, manufacturers, Jane's All the World's Aircraft 1966

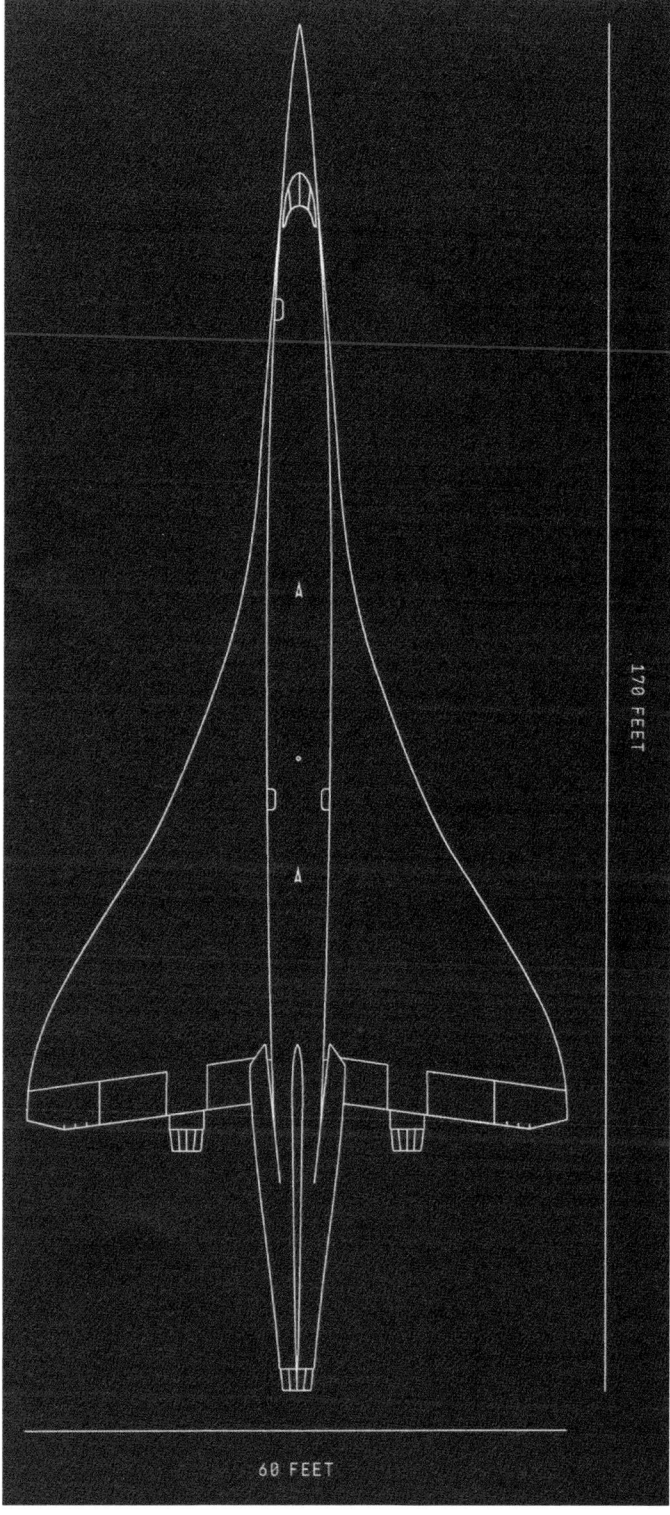

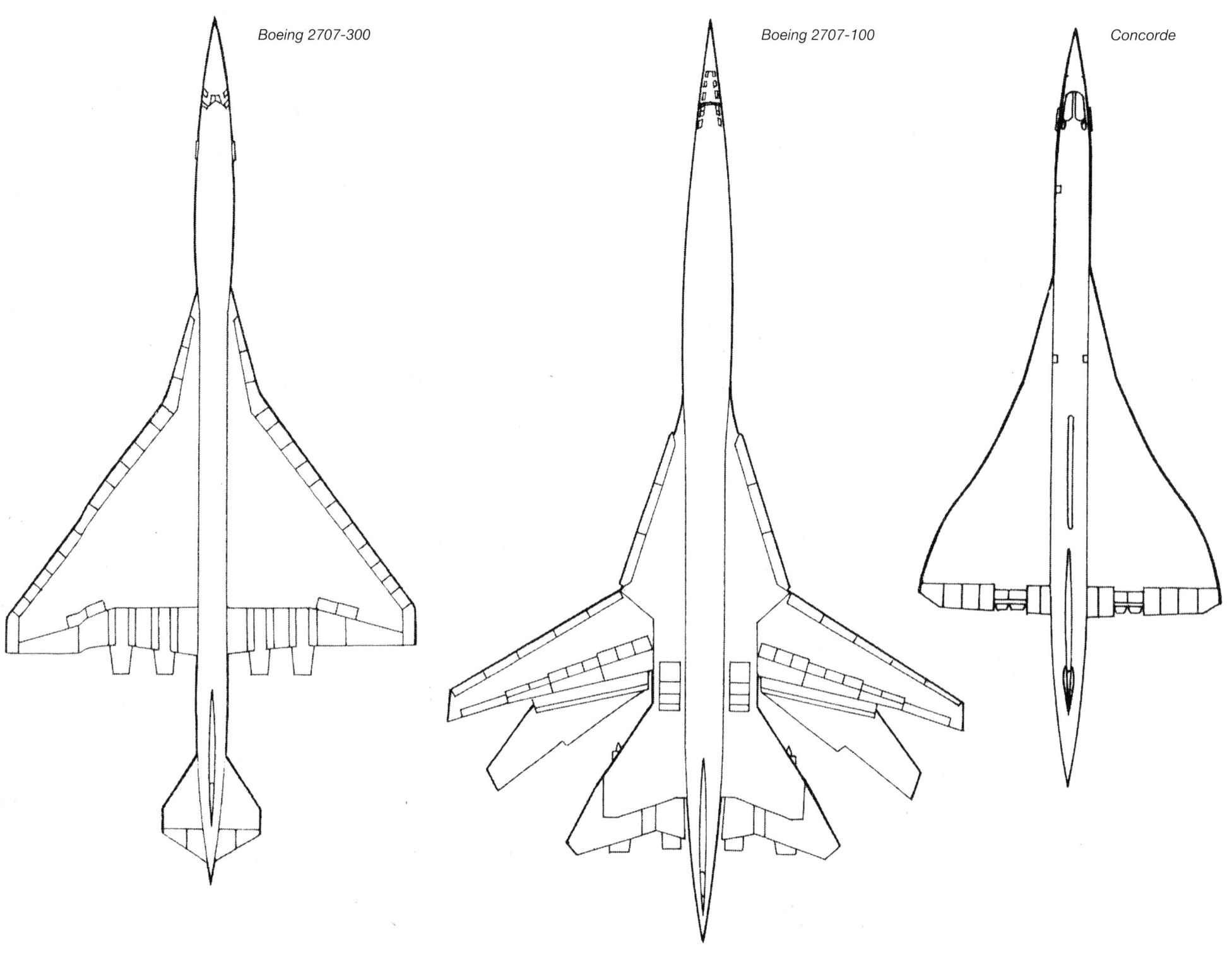

Boeing 2707-300

Boeing 2707-100

Concorde

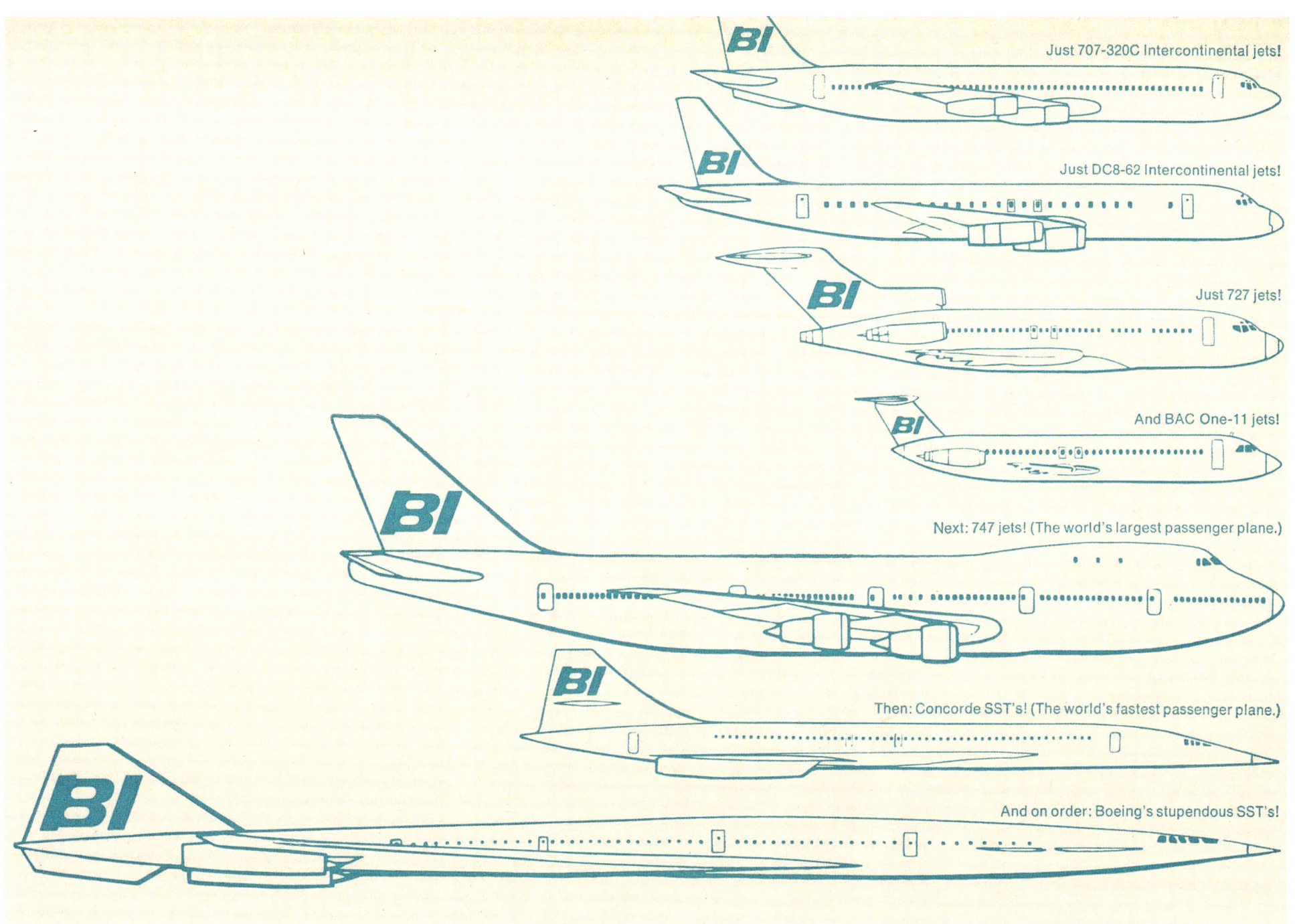

Just 707-320C Intercontinental jets!

Just DC8-62 Intercontinental jets!

Just 727 jets!

And BAC One-11 jets!

Next: 747 jets! (The world's largest passenger plane.)

Then: Concorde SST's! (The world's fastest passenger plane.)

And on order: Boeing's stupendous SST's!

Bibliography

Buttler, Tony with Jean Christophe Carbonel: Building Concorde. From Drawing Board to Mach 2, Manchester, UK 2018

Calvert, Brian: Flying Concorde, Shrewsbury, UK 1989

Chemel, Edouard: La vie du Concorde F-BVFB, Paris 2004

Chittum, Samme: Last Days of the Concorde – The Crash of Flight 4590 and the End of Supersonic Travel, Washington DC, USA 2018

Darlin, Kev: Concorde, Ramsbury, UK 2004

Davies, R.E.G.: Supersonic (Airliner) Non-Sense, McLean, VA, USA, 1998

Elser, Heinz: Concorde F-BVFB, Sinsheim, Germany 2004

Glancey, Jonathan: Concorde – The Rise and Fall of the Supersonic Airliner, London, UK, 2015

Gordon, Yefim and Rigmant, Vladimir: OKB Tupolev. A History of the Design Bureau and its Aircraft, Hinckley, UK 2005

Kreuzer, Helmut: Überschallverkehrsflugzeuge, Erding, Germany 2003

Lewis, Rob: Supersonic Secrets, London, UK 2003

Moon, Howard: Soviet SST. The technopolitics of the Tupolev-144, New York, USA 1989

Owen, Kenneth: Concorde and the Americans, Shrewsbury, London, UK 1997

Owen, Kenneth: Concorde – Story of a Supersonic Pioneer, London, UK 2001

Spaeth, Andreas: Concorde – Der Überschall-Passagierjet, Munich, Germany 2003

Trubshaw, Brian: „Concorde – The Inside Story", Stroud, UK, 2000

Audio documents

"Phone Calls: JFK is mad at Pan Am's Juan Trippe (June 4, 1963)"
https://www.youtube.com/watch?v=EuaZ0SkVf-Q&fbclid=IwAR1Ae1fAyJKg6xIP-v8YM0u7DSc5-W_ja8OlAzIb59Gx-xnPV5FNVgt9SD68

"Supersonic Travel with Blake Scholl, CEO of Boom Supersonic" 20. Mai 2020
https://ark-invest.com/podcast/fyi-ep66-supersonic-travel/

Text documents

Supersonic transport 1963. Memoranda from Najeeb E. Halaby, Administrator for the Federal Aviation Agency, to President John F. Kennedy regarding the potential construction of commercial supersonic transport (SST) aircraft
https://www.jfklibrary.org/asset-viewer/archives/JFKNSF/309/JFKNSF-309-001

Royal Aeronautical Society: "RAeS Concorde Conference", London April 8th, 2009, presentations and speech manuscripts

Websites

www.aerosociety.com
www.airlineratings.com
www.airspacemag.com
http://airwaysmag.com
http://atwonline.com
www.aviationweek.com
www.baaa-acro.com
www.bbc.com
www.cnn.com
www.concordesst.com
www.flightglobal.com
www.latimes.com
www.militaryfactory.com
www.mynorthwest.com
www.nasa.gov
www.nytimes.com

www.reuters.com
www.robbreport.com
www.seattletimes.com
http://simpleflying.com
www.spiegel.de
www.telegraph.co.uk
www.thenational.ae
www.tu144sst.com
https://en.wikipedia.org
https://de.wikipedia.org

Spike Aerospace, Boston MA, USA	
Lockheed Martin Skunk Works, Palmdale CA, USA	
Max Kingsley-Jones, London, UK	
Musée Air France, Paris, France	AF
Museum of Flight, Seattle, USA	
www.nasa.gov	
Tupolev PJSC Archive, Moscow, Russia	
Baz Genister	
Deutsche Lufthansa	LH

Institutions

Firmenarchiv Deutsche Lufthansa AG, Frankfurt

Publications

AIR International, UK
Aviation Week, USA
Esquire, UK
Flight International, UK
Flug Revue, Germany
Flying Review International, UK

Photo sources and abbreviations

Ad Jan Altevogt, Amsterdam, Netherlands	
Aerion Supersonic, Reno NV, USA	
Airbus Heritage, Toulouse, France	
Alexander Amelyushkin, Moscow, Russia	AA
Andreas Spaeth, Hamburg, Germany	AS
BAE Systems Heritage, Farnborough, UK	
Boeing Historical Archives, Seattle, USA via Archive	
Geoffrey Thomas, Perth WA, Australia	GT
Boom Supersonic, Denver CO, USA	
British Airways Heritage, London, UK	BA
Hanover Airport, Germany	
JACDEC, Jan-Arwed Richter, Hamburg, Germany	

A380
THE LAST GIANT

ARRIVING SOON
BY ANDREAS SPAETH

Flightcon International - Publishing Division

We are looking for our next publishing project

Specialists in Photo Books, Aviaiton Biographies (people, companies and aircraft)
Large Format, Military and Historical flight books

publishing@flightcon.net

www.flightcon.net